ENERGY EFFICIENT BUILDINGS

Edited by **Eng Hwa Yap**

Energy Efficient Buildings

http://dx.doi.org/10.5772/62592
Edited by Eng Hwa Yap

Contributors

Yousef Alhorr, Esam Elsarrag, Enes Yasa, Frank Barringer, Gary Moynihan, Dane Miller, Jeung-Hwan Doh, Barbara Matusiak, Abimbola Olukemi Windapo, Tülay Tıkansak Karadayı, İzzet Yüksek, Paulo Santos, Matjaz Prek

CBS, Edition **2018**

Published by InTech

Janeza Trdine 9, 51000 Rijeka, Croatia

Publishing Process Manager Martina Usljebrka

Technical Editor SPi Global

Cover Designer InTech Design Team

Additional hard copies can be obtained from orders@intechopen.com

Energy Efficient Buildings, Edited by Eng Hwa Yap
p. cm.
Print ISBN 978-953-51-2875-5
Online ISBN 978-953-51-2876-2

Contents

Preface

With an ever-rising world population, projected to hit 8.5 billion by 2030, compounded with the pressures of meeting goals to reduce CO_2 emissions and to effect environmental sustainability, the focus for energy efficient buildings has never been stronger. On top of that, the continuous drive for higher standards of living and a rise in the requirement for safe, comfortable and 'green' buildings to support our activities have resulted in an unprecedented demand for resources. These will have a knock-on effect upon buildings and the construction sector. For most nations, buildings and the construction sector represent a single largest activity, conventionally consuming energy in excess of 40% or more. This has resulted in its formidable share in greenhouse gas emissions.

This book aims to contribute to the on-going discourse on energy-efficient buildings and the role they play in a nation's effort to address climate change, energy consumption and greenhouse gas emissions, by considering buildings and the construction sector's unique position along a critical path to decarbonisation.

This book is divided into two parts—'Advances in Energy Efficient Building Design' and 'Building Energy Consumption, Demand and Efficiency', each part contributing to main theme and the on-going discussion from a multi-perspective and holistic viewpoint. Topics covered in the book range from daylighting, building topology comparison, building envelope design, zero-energy homes in hot arid regions, life-cycle considerations and energy efficiency analysis to managing energy demand through equipment selection. Each chapter addresses an important aspect of energy-efficient building and serves as a vital building block towards constructing a timely and relevant body of knowledge in energy-efficient buildings.

I hope that you will find this book useful and that the topics discussed are relevant. This book is a result of many months of hard work by all contributing authors, who have worked tirelessly and passionately. Without them, this work will not see the light of day. It is in this regard that I would like to acknowledge the professionalism and commitment from all chapter authors—thank you! I am also grateful to InTech for this opportunity and the Publishing Process Manager Ms. Martina Usljebrka for her wisdom, friendship, encouragement and all the help that she has offered throughout the process of editing this book—thank you!

Eng Hwa Yap
Associate Professor
University of Nottingham, Malaysia Campus
Malaysia

Section 1

Advances in Energy Efficient Building Designs

Chapter 1

Incorporating Sustainable Development Principles into Building Design

Jeung-Hwan Doh and Dane Miller

Additional information is available at the end of the chapter

http://dx.doi.org/10.5772/66825

Abstract

The main aim of this research is to utilise the focus of sustainable design to compare the material and subsequent environmental impacts of multi-storey structures situated in Australia. The structure types under investigation were characterised by post-tensioned and conventionally reinforced floor and roof flat plate slab systems. The foundation designs are undertaken for isolated spread footings on 32 structural model types with constant external dimensions which were composed of floor and roof slabs of varying concrete strength, span length and construction method, with all footing designs providing equivalent structural performance. The results from this study have reinforced the evidence that post-tensioned construction can have significant effects in reducing material requirements and provide increased structural and environmental efficiency. Through reducing the frame mass, the footing systems were able to be designed using significantly less embodied energy when compared to the reinforced concrete structures. It is also noted that further investigation in the foundational requirements of these models is warranted, with the need to investigate the use of mat foundations for cases where isolated spread footings have required more than 50% of the structural plan area and for the footings that have required excessively thick sections to resist large shearing actions for larger spanned cases at 10 and 13.33 m.

Keywords: embodied energy, sustainable development, sustainable design, global warming potential, post-tensioned structure, reinforced concrete structure, footing

1. Introduction

In recent times sustainable development and design has become an increasingly important issue to consider in our built environment [1–3]. The factors driving the adoption of sustainable development and design are numerous but perhaps the most significant is a growing concern about anthropogenic global warming caused through carbon emissions. The building and construction industry is a significant contributor to carbon emissions through the consumption of large quantities of energy. The main aim of this research is to utilise the focus of sustainable design to compare the material and subsequent environmental impacts of multi-storey structures in Australia.

Globally, greenhouse gas emissions are rising exponentially with most of this rise occurring in the last 60 years. Australia is ranked in the top 10 greenhouse gas emitting countries in the world with 25 t CO_{2-e} per capita in 2012 [4]. Australia's greenhouse gas emissions peaked in 2009. It has since declined, however now is not the time for complacency [1].

The introduction of energy efficiency provisions in the National Construction Code (NCC) [5] has attempted to address global warming issues by prescribing the requirements for operational energy efficiency. Soon after these changes the Green Building Council of Australia introduced the Green Star rating system which defines optional requirements above the NCC for energy efficiency and sustainability. Green Star's rise in popularity is a clear indication that building owners, developers and occupiers are demanding sustainable and energy-efficient buildings. Aside from addressing sustainability and global warming issues, stakeholders recognise the benefits that green buildings offer with lower operational costs being a prominent motivator.

The incorporation of the requirements of the NCC [5] and Green Star into buildings inevitably lowers operational energy consumption during a structures' life cycle. This has resulted in a more significant portion of the life cycle energy being represented by the structure itself which is referred to as embodied energy. This is clearly a target area for further reducing the energy consumed by a building [3]. Currently, the NCC [6] does not identify embodied energy as an area to improve the energy performance of buildings. Furthermore, the prominent building rating systems place insufficient emphasis on embodied energy considerations.

Due to the lack of emphasis placed on embodied energy outcomes, structural engineers have a limited obligation to incorporate measures requiring reduce energy consumption into the structural design. Typically design is governed by the requirements of the architect and classical building design objectives such as economy, utility, durability and comfort. Engineers should strive to not only achieve these requirements but to incorporate sustainability into their designs by improving structural efficiency and specification of appropriate materials. The building materials industry has recognised the importance of sustainability with extensive research being conducted in this area. The cement and steel manufacturing industries in Australia have adopted the use of alternative fuels and renewable energy sources in an attempt to lower greenhouse gas emissions and operational costs. The steel industry has also adopted the use of alternative manufacturing methods to preserve natural resources and utilise waste products. Likewise, the cement manufacturing industry uses waste products as partial cement replacements, often from the steel manufacturing industry, to preserve natural resources and lower operational costs [7].

The environmental impacts of the built environment and the need to incorporate sustainable design quantifying the environmental impacts of structures are continually evolving areas of research. This field would benefit greatly from increased industry collaboration.

2. Construction building material

The rate of human population growth globally has been exponential, and continued growth is seemingly inevitable. Australia's population alone is expected to increase to approximately 35.5 million before 2060; this growth comes with increasing pressures for infrastructure, housing and other related services [8]. This trend is unfortunately not limited to Australia, many other developed countries have similar predictions of population growth for the next 50 years, and the developing countries are showing trends of much faster population growth rates. These trends in growth are driving a demand that has ultimately led to the built environment becoming the largest single cause of anthropogenic climate change [9]. Construction, operations and maintenance of buildings have an estimated accumulation of 50% of all energy usage and as such are causing more than 50% of all anthropogenic greenhouse gas (GHG) emissions [10]. In Australia, approximately 30 million tonnes of finished building products are produced each year, with over 56% of this mass being attributed to concrete and 6% to steel [11] as shown in **Figure 1**.

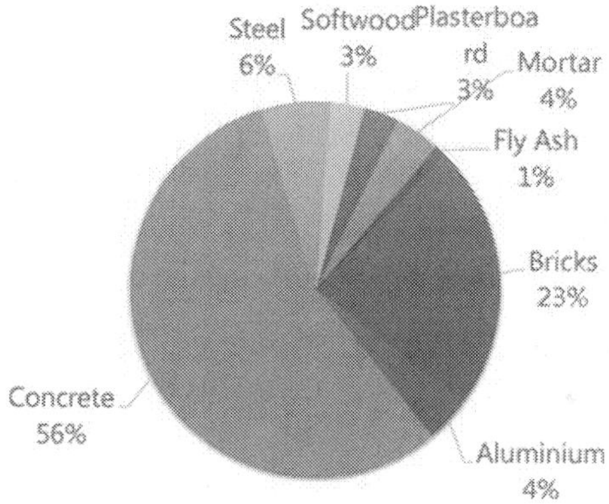

Figure 1. Finished building materials required in 2005, based on mass percentage. Image reproduced from the Australia Government [11].

Research undertaken in Australia also shows that concrete and steel are two of the highest contributing materials to global warming impacts at 12 and 29% for concrete and steel, respectively (**Figure 2**). This evidence underlines the importance of our management of these key building resources. Consequently, reducing the use of these materials in structures will show significant reductions in environmental impacts generated from the construction sector.

Once again this trend extends beyond Australia, with figures from other developed regions having similar percentages of concrete and steel making up the majority of finished building products. The major contributor to carbon pollution and the obvious associated environmental impacts in both the United States and Europe can be attributed to construction activities. The United States construction industry is responsible for using 40% of the country's total energy use, and 16% of the yearly water supply, additionally the industry consumes on an annual basis over 40% of the total mined raw stone, sand and gravel material and 25% of the logged raw timber [1].

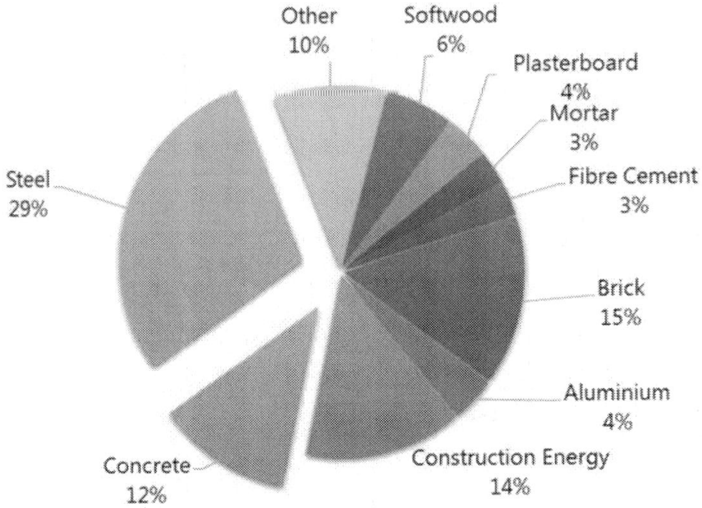

Figure 2. Global warming influence from construction materials in Australia [12, 13].

Countries participating in the Organisation for Economic Cooperative Development (OECD) demonstrated that 25–40% of the final energy consumption can be attributed to the building sector [14]. Having this knowledge enables a more complete understanding of the benefits and impacts that a sustainable construction (SC) approach will provide globally. As discussed the growing population, increasing demand for the built environment and its resources will continue to increase, therefore any improvements using more efficient approaches or technologies benefiting sustainability outcomes will be advantageous to the industry. When considering the magnitude of the global issues, even a small level of improved design efficiency could lead to significant reduction in negative environmental impacts. According to Hasegawa [14], by increasing the importance of improved efficiency in the structural design of buildings, two-thirds of the primary factors contributing to the poor environmental performance of building will be reduced, resulting in the minimisation of construction and demolition waste and in the reduction of CO_2 emissions.

3. Review of existing research

3.1. Post-tensioned and reinforced concrete suspended slab investigation

The use of post-tensioning systems on building construction is able to significantly reduce the concrete volume and steel mass required for a structure, resulting in substantial reductions in the structure's total weight [2]. As indicated in **Figures 3** and **4**, a reduction in concrete volume ranged between 5 and 23% as well as a reduction in steel mass ranging between 23 and 44%

http://dx.doi.org/10.5772/66825

is obtainable using post-tensioning methods as opposed to conventional reinforcing. The outcomes of the investigation indicated that not only are the post-tensioned office buildings more efficient in material usage, but more importantly in terms of an environmental impact assessment (EIA) through energy consumption and global warming potential.

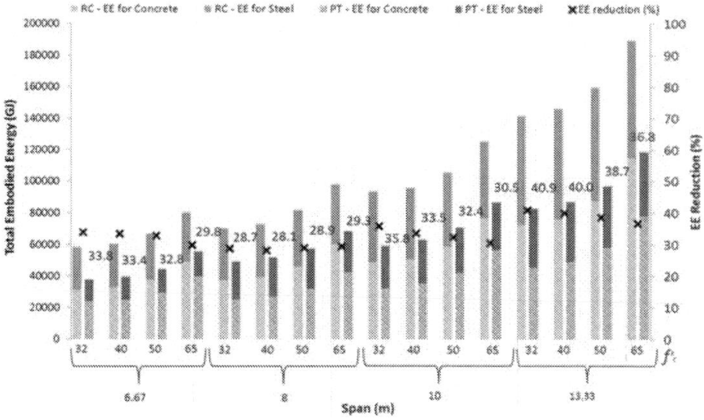

Figure 3. The required embodied energy values of various office structure types [2].

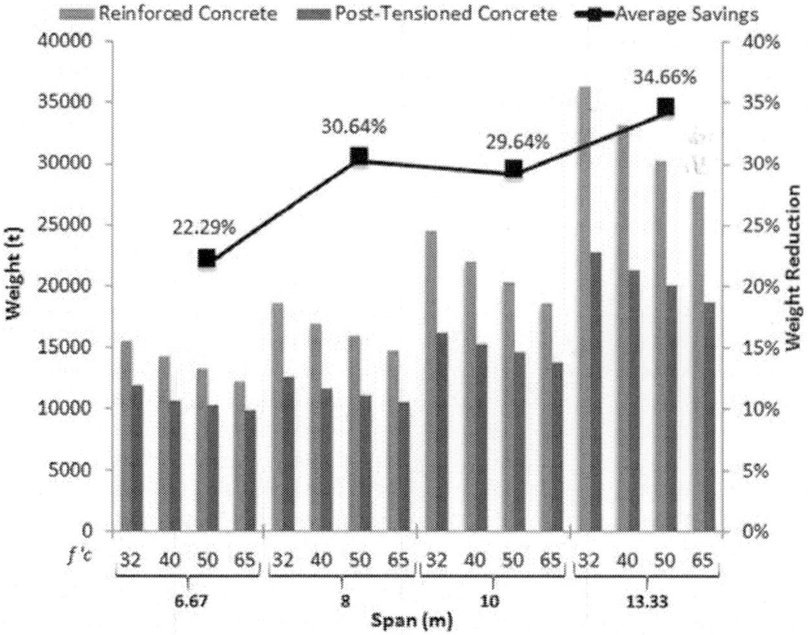

Figure 4. Variation in building weight by span length and concrete strength [2].

It was identified that the effectiveness of post-tensioning is greatest for spans exceeding 10 m, with the highest reduction in environmental impacts achieved being 41%. However, a minimum reduction in embodied energy (EE) and the global warming potential (GWP) of 28% was achieved for all structure types investigated [2].

Assessment of structural weight of slabs investigated was also undertaken. Results indicate structural weight savings for PT in comparison to RC construction. For the slab completing the required design task, achieving the least EE consumptions, average weight reductions between 22.3 and 34.7% were observed for PT buildings.

The disproportionate contribution of steel to EE of a structure compared to its contribution to structural weight was also investigated. Overall, the structure's weight accounted for between 40.8 and 59.6% of EE in previous study. These outcomes highlight the importance of steel usage optimisation for beneficial EE outcomes in concrete structural systems. Reduced structural frame weight can result in additional benefits to other structural components, including foundations, walls and columns that contribute to the structure's overall EE. Therefore, the overall reduction in EE of the entire structure through the use of PT and its overall effectiveness may be equal to or higher than those presented here. These additional considerations were outside the scope of this research but will be the focus of future investigations [2].

4. Environmental impact assessment (EIA) of footings

The main objectives of this study is to utilise the focus of sustainable design to compare the material and subsequent environmental impacts of footings for multi-storey structures situated in the South Eastern region of Queensland, Australia. The structure types under investigation were characterised by post-tensioned (PT) and conventionally reinforced (RC) floor and roof flat plate slab systems. The foundations footing designs were undertaken utilising isolated spread footings for the 32 structural model types. All the types had constant external dimensions which were composed of varying concrete strength, span length and construction method, with all footings designs providing equivalent structural performance. Following this, an environmental impact assessment (EIA) was undertaken, accounting for the embodied energy requirements for each varying structure type.

4.1. Methodology

The intention of this study is to utilise the process of structural design, to compare environmental impacts of mid-rise office structures in the South East Queensland region when PT and RC construction methods have been implemented. The results of this study shall further validate previous findings by Miller et al. [2], which have shown improvement of the environmental efficiency of office buildings where PT suspended floor and roof slabs have been used. Using embodied energy as an appropriate quantification tool for environmental performance, an evaluation of the structures foundation system shall be concluded. Specifically, the use of PT and RC raft footings for their respective PT and RC suspended floor system superstructures.

Previous research has been conducted that has investigated the potential reduction of environmental impacts. Previous research has investigated the potential reduction of the environmental impacts of concrete framed buildings. In particular, detailed comparisons were performed of variations in environmental performance between PT and RC buildings [1, 2]. Authors undertook an investigation into the effects of varying span lengths between columns and also increased concrete characteristic strength parameters, resulting in 32 structural variations (i.e. 16 post-tensioned and 16 reinforced concrete structural models each composed of 4 variations in span length and concrete strength). The current study is intended to carry on from the previous work conducted in this field and apply the methods of assessment to the foundations of the structures to investigate associated manifold benefits.

This study shall utilise the same identical model characteristics for the design of a raft footing option. Three case studies shall be investigated: **Case A** consists of an RC raft with an RC structure; **Case B**, an RC raft with a PT structure; and **Case C**, a PT raft with a PT structure. In total, 48 model variations of concrete building systems were considered.

Ultimately, this study will help identify the material savings that can be gained in the foundation system of a multi-storey structure and what outcomes can be achieved in terms of structural design efficiency in concrete buildings.

In order to achieve this, a multi-stage research methodology was formulated. This methodology was categorised into two major components, structural design and environmental analysis. The structural design involved several distinct components including: (1) design definition: including the formulation for the design of the specific building to be analysed along with the identification of assumptions necessary to undertake the analysis; (2) manual calculations in accordance with the Australian Standard [15] to provide a detailed design of the structural element varied, that was used for inputs into the two-dimensional computer analysis program, RAPT; (3) the structural designs were finalised using the results obtained from the computer analysis and these were verified using comparison with hand calculations to ensure accuracy and suitability of the design; and (4) the structural requirements for each element were subsequently detailed allowing a bill of quantities (BOQ) to be generated and an environmental performance assessment undertaken. The methodology has been summarised in **Figure 5**.

Utilising the allowable bearing capacity and applied working loads (1.0G + 1.0Q) initial size estimates (length and width) were determined in accordance with the allowable bearing capacity and checked with the inclusion of the self-weight of the footing.

An initial trial depth was utilised to check for the effects of two-way punching shear, an inadequate design sections resulted in an increase in the footings thickness. Once all the critical design cases were satisfied, detailing of the steel reinforcement was conducted to check for development length at critical sections and minimum reinforcing specifications as specified by the Australian Standard.

4.2. Environmental impact analysis

Reducing the material requirements of a structure and its elements is only one of a wide range of measures that have the capacity to effectively reduce the environmental impacts of the construction industry. In order to keep this project in line with previous research, the same unit measures of embodied energy have been applied to the current study. For the footings constructed of 40 MPa concrete, the correlation of embodied energy is as follows: for every cubic metre of concrete cast for the footings, 5670 MJ of energy is required [2]. For the embodied energy value of steel, 85.46 MJ is required per kg of installed reinforcement. These unit measures have then been applied to the material requirements, as determined from the bill of quantities, in order to determine environmental impacts of each footing system.

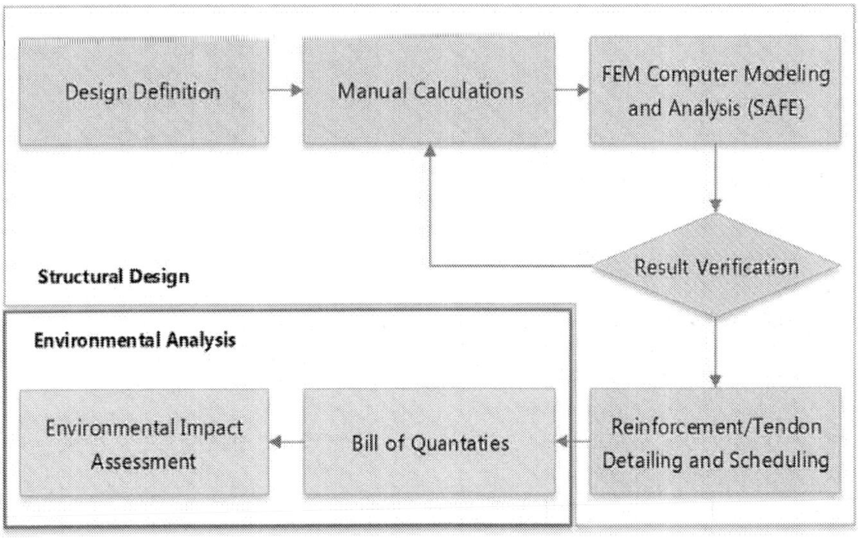

Figure 5. Methodology for structural design and environmental analysis.

In order to enable future widespread application of the presented research outcomes, it is intended that the environmental impacts be measured, be in the form of a relative estimation, EE. The results provided show an indicative comparison of the environmental impacts with the reduction in structural frame mass, associated with utilising PT slabs over RC slab frame elements. More importantly, the carry-on effects transmitted to the foundation of the structure allow for more lightweight footing systems to be utilised. These outcomes are presented in **Figures 6–11** and distinguish the individual contribution associated with concrete, steel and overall EE as well as the percentage reduction gained from utilising a favourable PT system.

4.3. Assumption

To undertake the design and analysis of each footing, the following assumptions were applied throughout the study:

- The soil conditions, i.e. soil bearing capacity and other associated characteristics are considered uniform throughout the site;
- Groundwater conditions are negligible and have no effect in terms of generating hydrostatic pressures on the base of the footing system;

- Overturning moment conditions generated from horizontal loading such as wind loads are negligible on the design actions of the footing system and will not be considered;
- Property boundaries pose no restriction upon the footing dimensions;
- The weight of overburden soil pressure can be neglected as it is assumed this has been accounted for in the allowable soil bearing capacity, based on geotechnical advice; and
- Shear reinforcement (stirrups) will not be utilised to resist one- or two-way shearing actions as in practice it is not considered economical to do so.

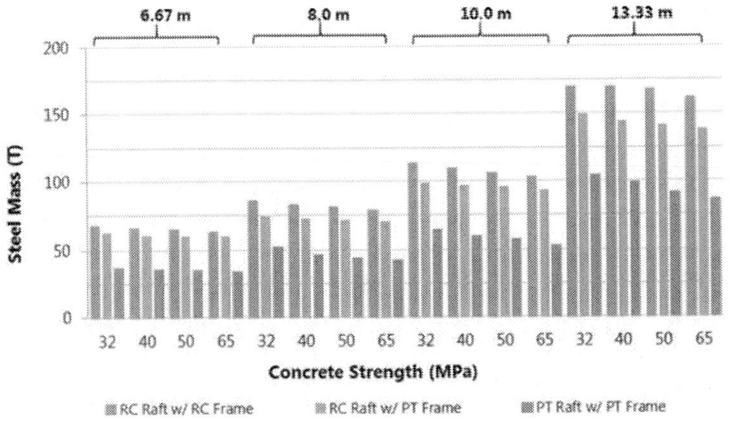

Figure 6. Comparison of concrete strength and span length with the steel mass required in each raft footing.

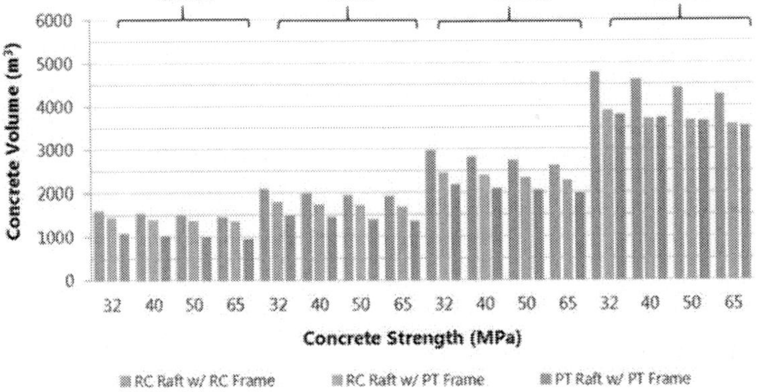

Figure 7. Comparison of concrete strength and span length with concrete volume in each raft footing.

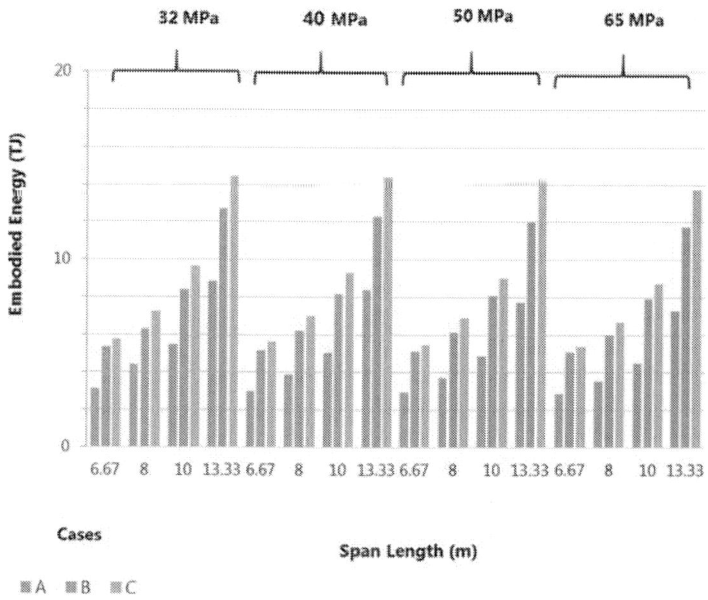

Figure 8. Overall trend comparison of embodied energy contributed by steel.

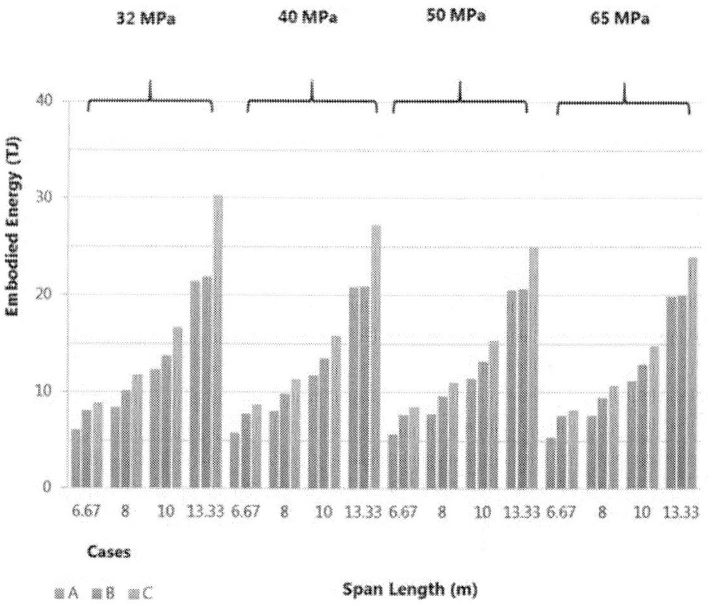

Figure 9. Overall trend comparison of embodied energy contributed by concrete.

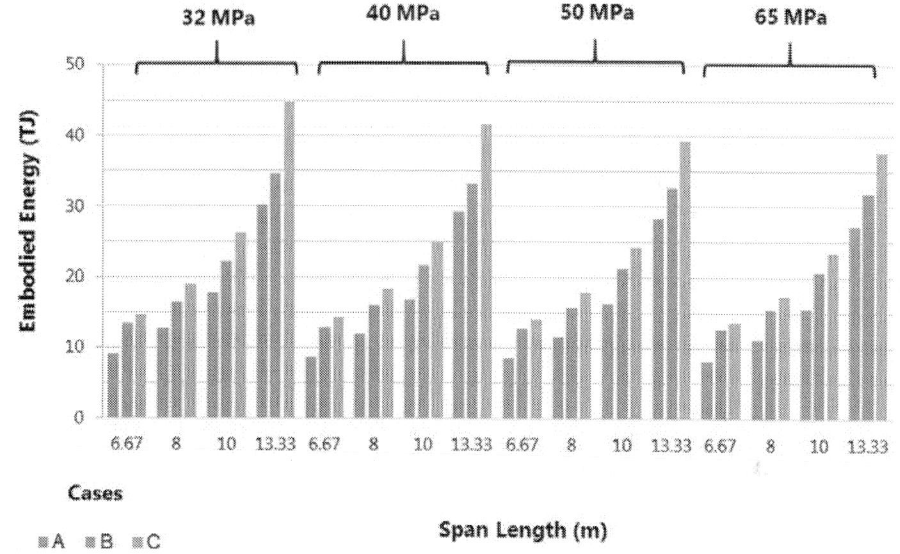

Figure 10. Overall comparison of embodied energy trend.

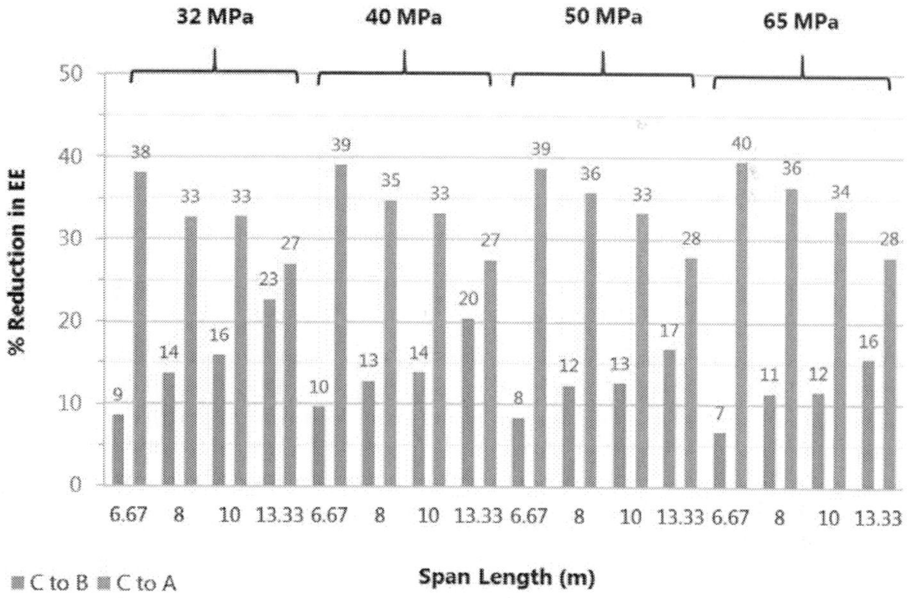

Figure 11. Percentage reduction of overall embodied energy between Cases C to B and C to A.

5. Results and discussion

As mentioned, the comparison of EE for each raft footing has been classified based on three cases based on the reinforcement/construction method utilised for the floor slabs and raft footing.

Case A: post-tension raft with post-tensioned structure

Considers a full PT system of floor slabs and raft footing solution. This case shall focus on examining the full effects of implementing a PT system for a foundation, specifically the benefits of utilising a PT raft.

Case B: reinforced concrete raft with post-tensioned structure

Focuses on a hybrid system of PT floor slabs and a RC raft footing system. This case shall specifically evaluate the efficiencies of reducing the structural frame mass by utilising PT in the floor slab system.

Case C: reinforced concrete raft with reinforced concrete structure

Focuses on a fully conventional system of both RC slabs and raft footing options. This case shall be the baseline for comparison of the efficiencies of PT alternatives, Cases A and B.

5.1. Effects of increased concrete strength and span distance

One of the main purposes of this study was to investigate the controlling effects of utilising increased concrete strengths for the suspended slabs of a structure, as well as increasing the span length. **Figures 6** and **7** present the results for both of these cases for the resultant quantities of steel and concrete individually.

5.2. Steel and concrete contribution to EE

The outcomes from the environmental impact assessment have been presented in **Figures 8–11**. The contribution to EE from steel is depicted in **Figure 8**. An increased span length shows a direct correlation to higher levels of EE. Case C continues to have the highest rates of EE, followed by Case B. Case A however appears to have a significantly lower EE level compared to both Case B and Case C. As the concrete strength increases no obvious change in trend appears for the EE levels in identical span lengths in any of the cases.

Figure 9 depicts the EE contributed by concrete with similar trends to **Figure 8**, with a direct correlation between span length and increasing embodied energy. Case C continues to have the highest rates of EE in all levels of span length, and Case B shows a significant reduction for the 13.33 m span length. For example, in 32 MPa concrete strength, Case B shows a 27% EE reduction compared to Case C. Case A continually shows the lowest EE levels in all span lengths. There is a noticeable decline in EE overall (in all cases and span lengths) as concrete strength increases.

Figure 10 shows the overall comparison of EE, which combines the outcomes from both **Figures 8** and **9**. This indicates that increased span length shows a direct correlation to higher levels of embodied energy. Also highlighting that Case C continues to have the highest rates of EE in all span lengths, followed by B and then A. **Figure 10** also shows a slight decline in EE overall (in all cases and span lengths) as concrete strength increases.

Figure 11 shows the percentage of reduction in overall EE between Cases A, B and C. The obvious trend is that the EE reduction percentage is significantly larger between C and A, than it is between C and B. For example, with a concrete strength of 50 MPa and a span length of 6.67 m, C to A has a 31% higher reduction in EE than the corresponding C to B comparison. The greatest savings for C to A can continually be seen in the smaller span lengths, with savings decreasing as the span length increases.

A clear trend was observed in **Figure 9**, comparing reductions from Cases C to B and from C to A. For C to B, as span length increases, the overall reduction percentage increases; however, this is opposite to the reduction trend in C to A, which decreases in EE reduction as span increases. The explanation for this is that Case A utilises a PT raft solution and that at smaller spans the use of PT contributes to a significant reduction in concrete volume and ultimately a higher EE reduction percentage; however, as span increases and the amount of columns decreases, the effectiveness of PT to add additional support to punching shear is reduced. The reasoning behind this is that, only a finite quantity of tendons can contribute to the resistance of two-way shear around the critical shear perimeter, this is due to the impracticality of fitting tendons so close to one another. Therefore, it is more favourable to have a larger number of columns to distribute the tendons between in order to increase punching shear resistance under each column.

Overall, this study has focused on quantifying the material saving and the associated EE through the utilisation of PT, it can be seen for Cases A and B, where PT has been utilised, that significant savings can be achieved. This was observed through both a hybrid PT option or a combined raft and floor slab system. The savings typically range between 7 and 40% at 6.67 m spans and 65 MPa concrete strength.

When considering the most efficient structure type in terms of EE and materials savings, it can be clearly recognised that a full PT raft and floor slab combination with 6.67 m span is the most effective.

6. Conclusion

As global population growth continues to increase into the future at an ever increasing rate, the impacts and draining effect on natural resources and the environment will continue. This work presents a strong discussion, which shows that a level of inefficiency currently exists in structural engineering design and environmental performance and that this should be rectified. As engineers, we can begin by bringing concepts of sustainable design into everyday practice. This investigation has delivered a comprehensive overview of the current state of knowledge in relation to the concrete design and construction industry, with the possible benefits achievable by improving its efficiency.

It was highlighted that structural engineers and other engineers alike, play a restricted role in the integration of sustainable alternatives to any given project, as they are at the mercy of the client and architects' demands. This shows that greater focus needs to be placed on the integration of sustainable design solutions into structural designs through aiming to improve the structural effectiveness of a building. This is achieved through improved efficiency of a structural system while offering adequate structural performance.

It has been shown in this investigation that through reducing material requirements in the frame of a structure, subsequent flow-on effects can be achieved in the foundation system when raft footings are utilised. Through reducing the mass of the frame, a direct correlation can be seen in the reduction of the applied loads needing to be supported by the underlying foundational system. For this investigation, the use of raft footings has been effectively designed and detailed for the 48 office structure configurations considered. These configurations were separated into three distinct cases. Case A consisted of an RC raft with an RC structure, Case B, an RC raft with a PT structure and Case C, a PT raft with a PT structure, all together totalling 48 model variations.

Clear findings have been presented showing significant material efficiencies achievable in the Case A footing system characterised by a PT raft/frame combination, followed by a Case B, a hybrid RC raft/PT frame option. Overall, the PT footing system outperformed the RC footing systems in all cases in terms of material reductions of steel and concrete and the EIA criteria for embodied energy. For the PT raft coupled with a PT slab system a significant saving ranging between 27 and 40% was achieved. This was predicted due to a distinct variation in column loadings generated by variations in slab element types and characteristics and due to the advantages offered from the contribution of pre-stressing methods. In terms of the effects of increased concrete strength and span length, results suggested that for material requirements, the PT raft/PT frame option outperformed all other cases and was more effective at taking advantage of the higher strength concrete. Overall, however, as the span increased, the overall requirements of the footing increased linearly; also, as the concrete strength increased, the material and EE requirements decreased linearly, with the highest saving found in the PT raft/PT slab option. When comparing the results for embodied energy to previous studies, it showed that spread footings for shorter spans resulted in less embodied energy than the PT raft solution. This however was shown to be negated, due to the considerations of constructability when spread footings acquire greater than 50% of the structural area.

Acknowledgements

The authors are grateful to Messrs, Robert Bartus and Mitchell Mulvey for their valuable work in the case studies. The authors also express to acknowledge an in-kind contribution from Mr. Tim Peters by providing guidance and computer software.

Author details

Jeung-Hwan Doh[1*] and Dane Miller[2]

*Address all correspondence to: j.doh@griffith.edu.au

1 Griffith School of Engineering, Griffith University, Gold Coast Campus, Queensland, Australia

2 Faculty of Society and Design, Bond University, Varsity Lakes, Gold Coast Campus, Queensland, Australia

References

[1] Miller D, Doh JH: Incorporating sustainable development principles into building design: a review from a structural perspective including case study. The Structural Design of Tall and Special Buildings. 2015;**24**:421–439.

[2] Miller D, Doh JH, Mulvey M: Concrete slab comparison and embodied energy optimisation for alternate design and construction techniques. Construction and Building Materials. 2015;**80**:329–338.

[3] Miller D, Doh JH, Panuwatwanich K, Oers van N: The contribution of structural design to green building rating systems: an industry perspective and comparison of life cycle energy considerations. Sustainable Cities and Society. 2015;**16**:39–48.

[4] Department for Innovation, Industry, Science and Research, (DIISR) (2009). Powering Ideas: An Innovation Agenda for the 21st Century. Canberra, Commonwealth of Australia.

[5] Australian Building Codes Board: National Construction Code (NCC) Building Code of Australia; Canberra, ACT; 2013.

[6] Australian Building Codes Board: National Construction Code (NCC) Building Code of Australia; Canberra, ACT; 2016.

[7] Wan Omar S, Doh JH, Panuwatwanich K, Miller D: Assessment of the embodied carbon in precast concrete wall panels using a hybrid life cycle assessment approach in Malaysia. Applied Mechanics and Materials. 2014;**10**:101–111.

[8] Australian Bureau of Statistics (ABS), (2010) 1370.0 - Measures of Australia's Progress, Viewed 17 March, 2012 from http://www.abs.gov.au/ausstats/abs@.nsf/Lookup/by%20 Subject/1370.0~2010~Chapter~Population%20projections%20(3.4).

[9] Intergovernmental Panel on Climate Change (IPCC), *GEO 4:* Environment for Development; United Nations Environment Programme (UNEP), Malta; 2007.

[10] Wood A: Sustainability: a new high-rise vernacular? The Structural Design of Tall and Special Buildings. 2007:**16**;401–410.

[11] Australian Government: Scoping Study to Investigate Measures for Improving the Environmental Sustainability of Building Materials, Canberra; 2006. Commonwealth of Australia.

[12] Walker-Morison, A., Grant, T. and McAlister, S. (2007). "The Environmental Impact of Building Materials". *Environment Design Guide* PRO 7.

[13] Walker-Morrison, A., Grant, T. and McAlister, S. (2007). "Strategies and Resources for Material Selection". *Environment Design Guide* PRO 8.

[14] Hasegawa T: Environmentally Sustainable Buildings: Challenges and Policies; OECD Publishing; 2003, Paris.

[15] Standards Australia: AS 3600-2009. Concrete Structures (Incorporating Amendment Nos 1 and 2); SAI Global Limited, Sydney; 2009.

Chapter 2

Daylighting is More than an Energy Saving Issue

Barbara Szybinska Matusiak

Additional information is available at the end of the chapter

http://dx.doi.org/10.5772/65866

Abstract

The focus in this chapter is to find adequate solutions to optimize daylighting of buildings. The climatic context is limited to Nordic European countries. As the knowledge of the positive impact of daylight on the human health increases and people spend most of their lives in buildings, the main objective is to create interiors with an optimum level of daylight for humans in a way which may also contribute to energy savings.

The solution for this complex task is showcased at a student's studio at the campus of the Norwegian University of Science and Technology (NTNU) in Trondheim, Norway. The studio has been refurbished in 2016 by fundamentally changing design of the skylights. A high potential for energy saving for electrical light has been created since the light level has been increased two to three times, both in cloudy and sunny conditions. The light is evenly distributed creating good conditions for visual tasks. Additionally, the change of the colour of light from yellowish to white makes that the room appears as brighter, cleaner, more spacious and more open.

The new design was developed in the frame of the DayLighting project carried out by the Light & Colour Group at NTNU.

Keywords: daylight, sunlight, skylight, daylight factor, view out, aesthetical perception

1. Introduction

For many years the focus in daylighting research has been at daylight level and daylight distribution in buildings. By adjusting these two parameters with a control system of electrical light and shading devices, a substantial amount of savings in energy consumption can be made possible. But the subject of daylighting also includes other issues like: view out, visual comfort and accessibility of sunlight in interiors as important parameters of occupant's health, well-being and aesthetical satisfaction.

1.1. Daylight for health

Research shows that daylight is advantageous (vs. electrical light) for visual tasks if it is delivered in a comfortable manner. It may create even and intense illumination enabling ideal conditions for visual tasks such as reading. Daylight is also the best light source for colour discrimination. The variation of daylight during the day stimulates the visual system and conveys information about weather fluctuation and the passing of time. The view out through the window is important for orientation and for keeping in contact with the outdoor environment; as such it is one of factors influencing the mental health of occupants.

Daylight is also crucial for non-visual neural system that is stimulated by the light falling at the special type of light-sensitive neural cells in the retina called intrinsically photosensitive retinal ganglion cells (ipRGC) [1]. The nerve impulses conveying information about light detected by those cells contribute to regulation of hormone production and to adjustment of human circadian rhythm to the 24 h cycle. The circadian rhythm is decisive for good functioning of our body and brain. Daylight is the optimal light source for stimulation of the non-visual system since it has a very high intensity and a spectrum containing much light in the blue-green part of the visual spectrum to which the non-visual system is most sensitive. As the morning daylight exposure of ipRGCs is most effective for the circadian rhythm adjustment, high daylighting level in rooms where we stay in the morning, as e.g. bedrooms, is especially important. Finally, we may underscore that importance of daylight for the health of occupants should be considered including both, visual and non-visual systems, which are two separate nerve paths, both receiving impulses from the retina in the eye but processing them differently in the brain.

As the adequate light exposure during the day is important for sleep length and sleep quality, and adequate duration of sleep is important for cognitive activity as, e.g. learning, it is logical to expect that adequately daylit buildings may contribute to better performance of occupants. Such correlations were actually found.

1.2. Local climate

Considering daylighting solutions the local climate has to be taken into consideration, this topic is limited here to the region encompassing Nordic European countries.

Let us start with the most important parameter that has impact at other parameters, i.e. the position of the sun. It may be observed that three national capitals and other large towns in northern Europe have latitudes close to 60°: Oslo 59°54′, Stockholm 59°19′, Helsinki 60°10′, Bergen 60°22′ and Orebro 59°16′.

Even a momentary look at the sun diagram for one of those cities, e.g. Oslo, and the sun diagram for, say, Cairo, helps to find the most fundamental difference between daylight in the North and in the South, i.e. the prevailing height of the sun over the horizon. The sun moves straight

http://dx.doi.org/10.5772/65866

up after sunrise in Cairo. In Oslo the movement is more horizontal; the elevation angle of the sun increases slowly over many hours and never reaches the area around the zenith. The highest position of the sun during the year in Oslo is 53.53°, the elevation angle at noon at the equinox is only 30.52°, while the two respective angles for Cairo are 83.37° and 60.33°.

Another interesting aspect is related to timing of the solar elevation angle. Since the position of the sun in Nordic countries is very low, the sun is near the horizon considerably longer than in countries situated at lower latitudes as, e.g. Cairo, 30°03′N. The very interesting question is how long can we expect the sun to be, e.g. between 0° and 10° above the horizon? The calculations of the percentage of daytime occurring during the first part of the year when the elevation angle of the sun is in intervals—(0°–10°), (10° –30°) and (over 30°)—were made with the help of the Solar Beam software [2] for Trondheim: 63°26′, see **Figure 1**. The results are also presented in **Table 1**.

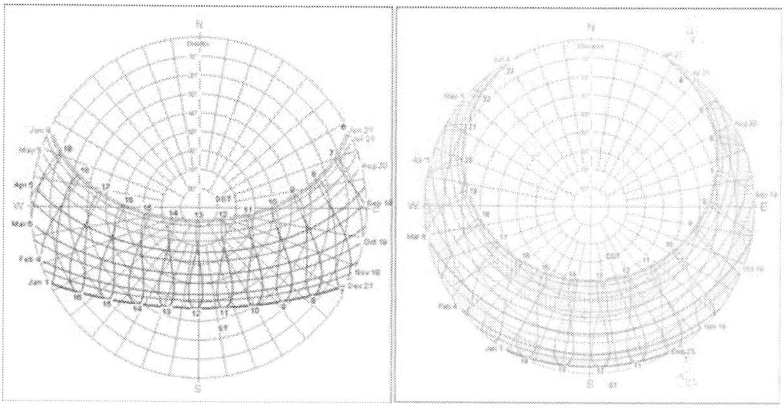

Figure 1. The sun diagram for Cairo and Trondheim generated using the Solar Beam software. The area of the Trondheim diagram representing 0–10° elevation angles of the sun is marked with red and the area representing elevation angles over 30° with yellow colour.

The highest elevation angle of the sun during the shortest day of the year, 21st of December, is only 3,35°. In the period from the 21st of December to the beginning of February the elevation angle of the sun will never reach 10°, this occurs first on the 3rd of February. After this day the number of hours when the sun is over 10° increases rapidly on a daily basis, but 30° sun elevation angle cannot be observed before the 30th of March.

The highest position of the sun in Trondheim is 50.01°. There are only four days during the year, 19th, 20th, 21st and 22nd of June, when the elevation angle of the sun is slightly higher than 50°; at 21st of June it lasts for 12 min.

The results are striking: the percentage of time during the year when the sun is between 0° and 10° is 35%. This means that for more than 1/3 of the whole daytime during the year we may expect a nearly horizontal light from the sun. The time when the sun is in the 0–10° angle interval is actually much longer than the time when it is over 30° (26%).

	0 10°	10–30°	30–50°	0–50°	Twilight
December 21–31	50.1	0	0	50.1	27.2
January	179.9	0	0	179.9	66.5
February	134.0	108.2	0	242.1	47.6
March	104.0	257.9	2.8	364.7	48.8
April	100.0	202.9	146.8	449.7	55.3
May	124.7	190.4	248.3	563.4	106.8
June 01–21	102.4	131.4	191.8	425.6	78.40
Together:	794.9	890.8	589.7	2275.5	430.42
	35%	**39%**	**26%**	**100%**	**19%**

Table 1. The time duration in hours when the sun is in the three elevation angle intervals, calculated monthly for the first part of the year in Trondheim.

The availability of the sunlight is also strongly dependent on the cloud cover. To look at the frequency of sunny skies a map was generated with the help of the Satel-Light [3], see **Figure 2**.

The analysis made so far shows the typical features of daylight in the Nordic countries:

1. Dominating low solar elevation angle during the year.
2. Long periods of twilight and white nights in the time period close to summer solstice, midnight sun at places north of the Arctic Circle.
3. Rather low frequency of sunny skies during the year, especially during winter.

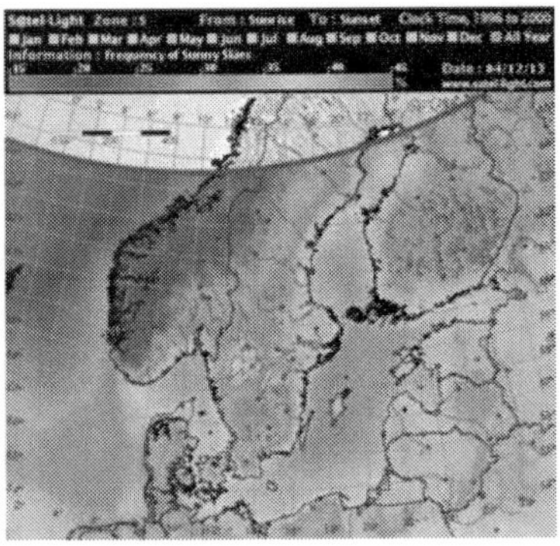

Figure 2. Frequency of sunny skies in North Europe.

http://dx.doi.org/10.5772/65866

2. How to utilize daylight?

A new method of harvesting daylight in the Nordic climate was developed during the retrofitting of a studio located at the campus of the Norwegian University of Science and Technology in Trondheim, Norway. The room is located on the third floor of a three storey building called Lavblokk Sør. The building is situated immediately south of a high raise 13 storey building, representing its largest exterior obstruction (**Figure 3**).

Figure 3. NTNU campus. The studio is marked by a red rectangle within the red ring.

The room is 8.2 m × 14.3 m, a total of 117 m^2 and has a height of 3.4 m, as shown in **Figures 4** and **5**. It is one of eight student studios at the Faculty of Architecture. The students use them as permanent working places; the typical visual tasks are drawing, making sketch models and of course reading and writing both in the analogue and digital format.

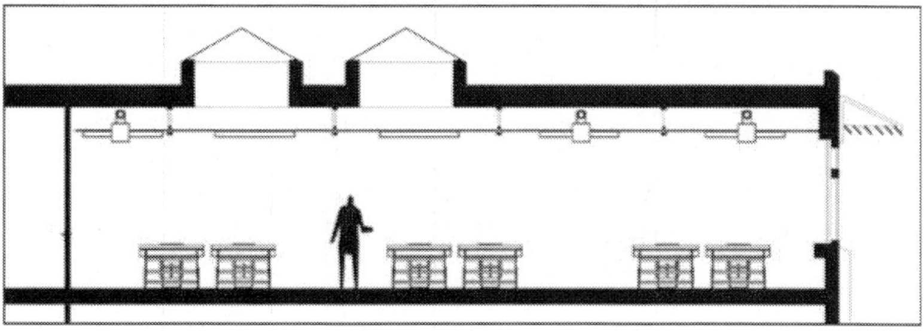

Figure 4. The cross-section of the room.

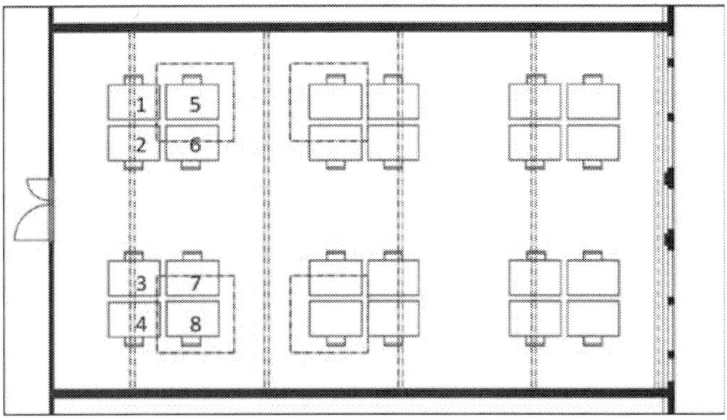

Figure 5. The layout of the room with drawing tables for 24 students. Location of skylights is marked with dashed squares. The dashed lines represent steel beams.

The roof construction is made of steel beams situated across the room, as shown in **Figures 4** and **5**. There are four skylights in the rear part of the room. The original idea behind skylights was to deliver daylight in the rear part of the room and to protect against direct sunlight falling at working areas. This was achieved by the usage of translucent acrylic sandwich elements in the pyramid-shaped skylights. Skylights are square in plan 1.8 m × 1.8 m; the height of vertical walls surrounding each roof opening, the well, is 0.85 m.

The acrylic sandwich elements diffuse sunlight very well, but they have also a negative feature, namely, the total transmittance of light is low, i.e. sandwich elements transmittance is 30–40% depending on the thickness of material, something that limits daylight penetration during cloudy weather which, as was shown in introduction, occurs in Trondheim quite often.

After including this refurbishment project in the research activity of the Light & Colour Group, specifically in the DayLighting project carried in international cooperation, in the frame of IEA Task 50 "Advanced solutions for retrofitting of lighting systems" [4], the scope of the project has been widely extended.

The main objectives of the refurbishment project were to:

- Keep the energy consumption for heating at the same or lower level. It should be mentioned that since the climate in Trondheim is rather cold and a high rise building is situated to the south, there was no need for cooling in the studio.
- Maximize the provision of daylight, both sunlight and diffuse light from the sky; this is to enable a reduction of energy consumption for electrical lighting.
- Create good visual conditions for all occupants in the room in any daylight condition.

The scope of the project was originally limited to replacement of dirty old acrylic sandwich elements with two-layers low-energy glass (**Figure 6**).

Figure 6. Existing skylights from the outside and from the inside.

To minimize the energy consumption for heating three-layers low-energy glass was chosen instead of two-layers such that the U-value of the new glazing, specified at the middle of the glass, has been 0.9 W/m^{2}K.

The estimated transmittance of old sandwich elements is in the range 20–40% depending on the dirt thickness and age (yellowing of material) while the transmittance of the new glazing oscillates around 70% depending on maintenance frequency. To maximize the daylight provision the low-iron glass was chosen to keep the light transmittance of the glazing at the topmost level (about 4% increase comparing to the standard glass).

The vertical well walls have been covered with highly reflective mirrors, R = 96%. Such a solution ensures that all of the light hitting surfaces of the mirrors is reflected down, i.e. not a single ray is reflected back to the atmosphere. This is especially important in the location with dominating low sun.

The most challenging task was to ensure the visual comfort. The sunlight passing through the glazing and reflected from mirrors would create patches of strong sunlight in the interior. The patches would move in the room as the sun moves on the sky. The crucial question was how to distribute the sunlight in the room and at the same time keep the skylight transmittance at topmost level?

To obtain the most promising solution different alternatives were tested in a scale-model study carried out in the Daylight Lab. at NTNU, Light & Colour Group. The alternatives assumed utilization of perfectly transparent acrylic plates which have been perforated with holes made with the Laser Cutting Machine accessible at the workshop at the Faculty of architecture, NTNU.

After the idea of scattering sunlight with the help of perforated acrylic plates proved to be successful, a long design process followed. A suspension system for hanging the plates exactly at the level needed to be aligned with the ceiling plates had to be designed. The size of the acrylic plates had to be selected according to the dimensions of the skylight well. The thickness of the plates had to be chosen according to the size of the plates and the size of the holes and the distance between the holes had to be adjusted according to the thickness of the plates.

The result of this project is shown in **Figures 7** and **8**. Pictures were taken with a Nikon D600 camera with the fish eye EX DG positioned at the entrance to the room about 2.0 m above the floor. Two rooms were photographed, the renovated room (on the left-hand side) and the neighbouring room that has not been renovated (on the right-hand side). A series of 11 low-dynamic range pictures were taken and combined afterwards into high-dynamic range HDR pictures. The HDR pictures visualize a room in a way that is much more similar to the way human visual system perceives it, especially regarding lightness of surfaces.

It is clear that the general light level in the retrofitted room is significantly higher, both in the overcast and sunny conditions. This may be observed comparing the right and the left picture but also by looking at the high situated windows in the partition wall. The windows appear as darker than the wall if seen from the renovated room creating impression that we are looking into a darker room to the right; a contrary impression is created if the same windows are seen from the room which was not retrofitted.

Additionally, the renovated room appear as cleaner, since the yellowish shade had been removed. Interestingly, the brightest spots in the renovated room are at acrylic panels, even very gentle sun patches on side walls appear as darker.

Figure 7. Fish eye HDR pictures of the studio under overcast sky conditions, the new solution to the left, old one to the right.

Figure 8. Fish eye HDR pictures of the studio under clear sky conditions, the new solution to the left, old one to the right.

http://dx.doi.org/10.5772/65866

3. Potential for energy savings

The refurbishment project is not completely finished; the next step will be the replacement of the electrical light system. New LED-luminaires will be fixed underneath the steel beams and will be connected to a daylight-responsive control system. This will allow theoretically the highest possible energy saving for lighting.

The results of illuminance measurements taken in June 2016 at the middle of each numbered desk, see **Figure 5**, both in sunny and cloudy conditions are presented in **Table 2**. For cloudy conditions both illuminance and daylight factor (D) values are presented.

On average the light level is increased by two to three times both in sunny and cloudy conditions. We predict similar values for intermittent sky, i.e. more than doubling of the light level. The daylight factor was increased from 2.1–2.3% in the room with the old solution to 4.8–6.9%.

	Sunny		Cloudy			
	New (lx)	Old (lx)	New (lx)	D	Old (lx)	D
1	3145	1655	1306	6.3%	454	2.2%
2	3485	1795	992	4.8%	435	2.1%
3	3490	1760	1154	5.6%	433	2.1%
4	3110	1730	1417	6.9%	456	2.2%
5	3960	2015	1252	6.1%	472	2.3%
6	4300	1900	982	4.8%	462	2.2%
7	7535	1820	1109	5.4%	431	2.1%
8	4445	1765	1327	6.4%	458	2.2%

Table 2. Illuminance measured on eight desks situated closest to the entrance.

In sunny conditions, while the sun elevation angle is about 48°, the light level in the room with old skylights oscillates between 1600 and 2000 lx, in the room with the new solution it is around 3100–4400 lx. The variation in sunny conditions is higher as a very small percentage of light is allowed to pass directly through the acrylic panels creating a little brighter area (7535 lx).

As the light level in the room with old skylights is rather low, doubling of this value makes daylight illumination much more adequate to the needs of the users (300–3000 lx). This means that the time of the daylight level on desks higher than 300 lx has been radically increased, giving large potential for energy saving for lighting.

4. The importance of the view out of the window

The research conducted by Ne'eman et al. [5] shows that acoustic, heating, lighting, outside view, ventilation, air conditioning and the design of work space are among the most important physical environmental factors affecting worker satisfaction. Daylighting and outside view, both conveyed by windows, are the most important factors for achieving worker satisfaction. The view out is especially important for well-being since it gives the possibility to keep contact with the outside environment continuously [6, 7].

The existing hypotheses and research findings about preferences for view may be categorized into three groups: the need for information about the outside environment, the need for aesthetical experience and the need for restoration and health [8].

The need for visual information is well described by Lam [9]. A typical window conveys information about location, time and weather conditions, but also about activities and events outside the building. If the need for visual information is not satisfied, the ability to focus on a work task is seriously limited.

The need for aesthetical experience has been explained by the evolutionary aesthetics [10, 11] claiming that perceptions of beauty are evolutionarily determined, i.e. places and landscapes which people consider beautiful are typically found in settings that are likely to support survival of the human's genes. This theory together with the prospect-refuge theory proposed by Appleton [12] may explain why natural landscapes and well-kept buildings are generally preferred.

The need for restoration and health can be explained by the attention restoration theory [13]. Natural environments have qualities that in combination seldom occur in other types of environments. They give a feeling of being away and they create a sense of extend, i.e. what is seen is a part of a larger area. In addition, natural environments create a sense of fascination meaning that they encourage exploration, attract attention and hold it effortlessly [14, 15].

Consequently, for the evaluation of the view out, two categories of descriptors should be considered:

- the content of the outside view discussed previously, i.e. qualitative descriptors, and
- the extent to which the view is actually available from a given point in the interior, i.e. quantitative descriptors.

Both types have impact on the view quality, see **Tables 3** and **4**, but the responsibility may be allocated to different professional groups. Architects designing a single building have responsibility for most quantitative descriptors, as e.g. the width of the view from a given place in the building as well as the possibility to see the view layers: sky, landscape and the ground. To a certain degree they may decide about the view distance, e.g. by choosing the location of the building on the site. Additionally, architects may attempt to include a nice landscape element in the views from many rooms. However, there are urban and landscape planners and municipalities who decide about the distance between buildings, width of the streets, size of town squares and have responsibility for creation of attractive urban or rural landscapes, i.e. are decisive for the qualitative descriptors.

	View quality		
Descriptor	**Sufficient**	**Good**	**Excellent**
Width of the view (glass)	> 14°	> 28°	> 54°
Outside distance of the view	> 6 m	> 20 m	> 50 m
Number of view layers: - sky - Landscape (urban and/or nature) - Ground	At least landscape layer is included	Minimum two layers are included	All layers are included
Environmental information: - Location - Time - Weather - Nature - People	Time, weather and basic info about location	Time, weather, location and nature or people	All

Table 3. Assessment of the view out, quantitative descriptors.

	View quality		
Descriptor	**Sufficient**	**Good**	**Excellent**
Content and Quality of landscape: - urban low quality, as e.g. toward concrete walls, parking plot, etc. - urban middle quality - urban high quality, e.g. an attractive square or with attractive elements as, e.g. historical buildings, fountains, sculptures, etc. - natural/rural low quality, as e.g. monotonous rural landscape - natural/rural middle quality - natural/rural high quality, e.g. varied and including beautiful elements as lakes, well-shaped trees/bushes	Natural/rural –low or urban middle	Natural/rural -middle or urban high	Natural/rural -high or urban high
Composition: - poor balance between landscape elements, e.g. more than ½ of the view is toward one dominating element - good balance - very good balance between landscape elements	Poor balance but at least the central part of the view is free from undesired elements.	Good balance and the central part of the view is free from undesired elements.	Very good balance and the view is free from undesired elements

Table 4. Assessment of the view out, qualitative descriptors.

It is preferable to divide the view out quality into classes, e.g. sufficient, good and excellent, see **Table 3**. For a good view it is recommended that the width of the view (window glass only) from the observation place in the room is larger than 28°, the mean minimum outdoor distance to other buildings/constructions is minimum 20 m and minimum two of the three view layers are included. If the evaluation of those metrics is difficult, it may help to consider if the view conveys environmental information about time, weather, location and nature or people, see the last row in **Table 3**.

As discussed previously, for a view to meet the occupants' needs for environmental information, for aesthetical experience and for restoration there are expectations to the quality of the view. Interesting question is then, how people evaluate quality of the view from their usual occupancy places. In the research study performed in a previous work [8] in Trondheim, Norway, over 100 subjects were visited at their work places and asked to evaluate the quality of the view out. The quality was best predicted by the view distance, the number of view layers, the quality of the landscape/elements and the composition of the view. This is a new finding showing that the aesthetical experience may be more important than assumed previously. Occupants appreciate the possibility of looking towards beautiful and well-maintained buildings and towards well-kept parks and gardens. It appears also that a certain degree of complexity is preferred. A variation in composition of green areas with different shapes and/or colours of trees and shrubs is preferred. Regarding buildings, a certain variation of forms and shapes is preferred compared to ordinary plain walls made of one material.

Anyhow, even if the view contains a beautiful element, like for example a cathedral, the important question is how much of this element is actually included in the view? Another question is if there are unwanted elements, like e.g. a poorly maintained building, in the view? With other words, how is the composition of the view? Where the liked and disliked elements are positioned, at the central or peripheral part of the view? A good method to find it out is simply to take pictures from the place under consideration. **Table 4** gives additional guidance regarding the qualitative descriptors.

It has to be mentioned that it may be also indirect determinants that influence people's perception such as glare, colour rendering of glazing, contrast and clarity of the image which may have impact on the final evaluation of the quality of the view.

5. Daylight and aesthetics

Experience-based knowledge gathered by generations of architects and building planners tells us that daylight has a profound impact upon the aesthetic ambience of the interiors. However, can this be scientifically proven? A full-scale study with mock-up rooms build in the Room Lab at NTNU in 2014 [16] examined the impact of window size and room reflectance on the perceived quality of a small room, and the correlation between various architectural quality attributes, as shown in **Figure 9**.

Figure 9. Pictures of the mock-up rooms.

The mock-up rooms were visited in the randomized way by a number of carefully selected subjects. Statistical analysis of their answers led to the following conclusion: "overall, the rooms with larger windows and lighter walls obtained higher ratings for all the studied architectural quality attributes, suggesting that high levels of daylight are crucial in order to achieve a more pleasant, exciting, complex, legible, coherent, spacious, open and spatially defined room." Additionally, it turned out that the window size was more important for the aesthetical judgement than the surface colour; on average the black room with the largest window scored higher than the white room with a small window. As this study was limited to rather small rooms, achromatic colours and overcast sky conditions, more research is needed for generalization.

Another full-scale experiment was carried out in a small office room situated in one of high rise buildings on the NTNU campus to find out the aesthetical preferences of four different daylighting systems: venetian blinds (WB), high reflecting blinds (HRB), hybrid (included electrical light also) light shelf (HLS) and mirrored light shelf (MLS) [17], **Figure 10**. Results from MANOVA indicated that both the daylighting systems and the type of sky had an effect on the aesthetical attributes, and the significant interaction effect suggested that the aesthetical perception of daylighting systems depends on the type of sky. The room equipped with the high reflecting, i.e. specular, blind system, which created even and rather strong illumination, was evaluated highest by participants, i.e. as the most pleasant, exciting, coherent, spacious, legible and the one making the room most spatially defined, under both clear and overcast sky conditions.

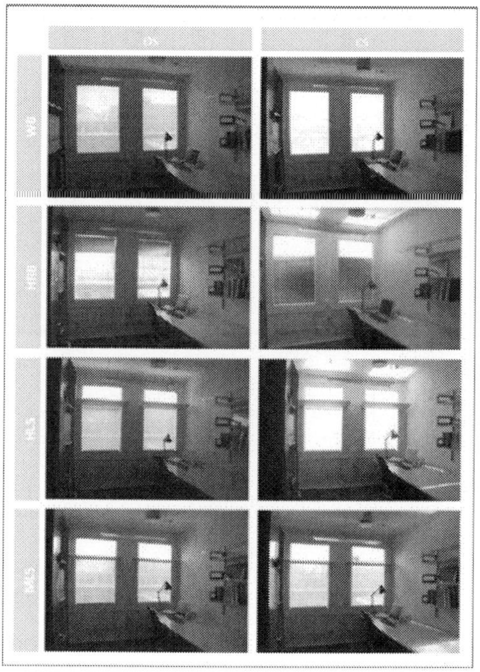

Figure 10. Pictures of the studied office rooms with the eight different stimuli.

6. Conclusion

The most prominent question in the study of daylighting is how to create rather high daylight level in cloudy conditions and avoid uncomfortable sunlight in sunny conditions in the same space. This is especially problematic when intense sunlight causes discomfort when shone on human bodies and distracts our view on computer screens. Sun shading devices are usually used to solve this problem, but this will cause the overall light level in the room to diminish and electrical light has to be used.

The new system developed by the Light & Colour Group at NTNU [18] in Trondheim is based on the strategy of maximal penetration of daylight through the skylight, aided by vertical mirrors on the walls of the skylight well, which is scattered in the room using perforated acrylic plates. The results observed showed evenly distributed light of high intensity providing comfortable working conditions and significant energy saving when the use of electrical lighting is reduced. For high sun elevation angles which happen at the middle of the day during summer, sun patches may occur, especially on the side wall, but then they are weak and does not cause discomfort. In fact, they add luminance contrasts in the room and in this way increase esthetical experience. As the studied room has rather large windows towards trees, the good view is also ensured.

Our research confirms that there is clear potential for energy saving for lighting by utilization of daylight directly for illumination of interiors using canny designed skylights.

Author details

Barbara Szybinska Matusiak

Address all correspondence to: barbara.matusiak@ntnu.no

Light & Colour Group, Norwegian University of Science and Technology (NTNU), Trondheim, Norway

References

[1] Lucas, R.J., et al. Regulation of the mammalian pineal by non-rod, non-cone, ocular photoreceptors. Science Vol. 284, Issue 5413, pp. 505–507, 1999.

[2] Solar Beam (2013) web-page: http://solarbeam.sourceforge.net/

[3] Satel-Light (2013) web-page: http://www.satel-light.com/core.htm

[4] IEA Task 50 (2016) web-page: http://task50.iea-shc.org/

[5] Ne'eman, E., Sweitzer, G., Vine, E. (1984) Office worker response to lighting and daylighting issues in workspace environments: A pilot survey. Energy and Buildings, 6(2): 159–171.

[6] Christoffersen, J., Johnsen, K. (2000) Daylighting and window design. Post occupancy studies in environments. Light & Lighting, 19, 31–33.

[7] Veitch, J.A., Galasiu, A.D. (2012) The Physiological and Psychological Effects of Windows, Daylight and View at Home: Review and Research Agenda NRC-Construction, IRC-RR-325, Canada.

[8] Matusiak, B., Klöckner, C. (2015) How we evaluate the view out through the window. Architectural Science Review. Published online: 29 Apr 2015.

[9] Lam, W.M.C. (1977) Perception and Lighting as Form Givers for Architecture. New York McGraw-Hill, Inc. 1977.

[10] Dutton, D. (2003) Aesthetics and Evolutionary Psychology in "The Oxford Handbook for Aesthetics". Oxford University Press.

[11] Dutton, D. (2009) The Art Instinct: Beauty, Pleasure, & Human Evolution. New York, Oxford University Press, ISBN: 978-0-19-953942-0.

[12] Appleton, J. (1975) The Experience of Landscape. London: John Wiley.

[13] Kaplan, S. (1987) Aesthetics, affect and cognition: Environmental preference from an evolutionary perspective. Environment and Behavior, 19(1):3–32

[14] Kaplan, R., Kaplan, S. (1989) The Experience of Nature. Cambridge: Cambridge University Press.

[15] Kaplan, R., Kaplan S., Ryan, R.L. (1998) With People in Mind: Design and Management of Everyday Nature. Washington, D.C: Island Press.

[16] Moscoso, C., Matusiak, B., Svensson, U. P. (2015) Impact of window size and room reflectance on the perceived quality of a room. Journal of Architectural and Planning Research. Vol 11(4).

[17] Moscoso, C., Matusiak, B. (2015) From windows to daylighting systems: How daylight affects the aesthetic perception of architecture. Proceedings of the CIE-2015.

[18] Light & Colour Group, web-page: http://www.ntnu.edu/bff/lightandcolour

Chapter 3

Energy Efficiency of Lightweight Steel-Framed Buildings

Paulo Santos

Additional information is available at the end of the chapter

http://dx.doi.org/10.5772/66136

Abstract

The market share of lightweight steel-framed (LSF) construction system has grown over the last decades, mainly in low-rise residential buildings, due to its advantages such as having small weight with high mechanical strength; reduced disruption on-site and speed of construction; great potential for recycling and reuse; high architectural flexibility for retrofitting purposes; easy prefabrication, allowing modular construction; economy in transportation and handling; superior quality given off-site manufacture control; and excellent stability of shape in case of humidity and resistance to insect damage. However, given the high thermal conductivity of steel and the lightness of this type of construction, it may also have some drawbacks if not well designed and executed. Therefore, special attention should be given to the LSF building envelope in order to minimize thermal bridges. Moreover, given the usual reduced thermal mass, several strategies could be implemented to increase thermal inertia, consequently reduce indoor temperature fluctuations, enhance the occupants comfort and increase energy efficiency. In this chapter an overview of the main features related to the thermal behaviour and energy efficiency of LSF buildings is provided alongside some related case studies.

Keywords: LSF buildings, energy efficiency, thermal behavior, thermal bridges, thermal inertia, case studies

1. Introduction

Sustainable development and energy efficiency are two of the most relevant concerns of today's humankind. Therefore, the demand to reduce energy consumption and to use more environmental friendly materials is increasing. In fact, today there is no doubt about the link between the burning of fossil fuels and the consequent release of carbon dioxide with climatic changes, for example, global warming and extreme climate events. Buildings exhibit an enormous potential to mitigate greenhouse gas (GHG) emissions when compared with other activity sectors [1]. Thus, the use of renewable energy sources (RES) and the reduction of energy consumption are two top priorities, being these the major challenges of the twenty-first century for emerging and developed countries.

In this context, several objectives were established by the European Union in the Energy Performance Building Directive—EPBD (European Directive 2010/31/EU [2]) regarding "nearly zero-energy buildings" for the year 2020. The EPBD addressed both the increase in RES and the improvement in buildings energy efficiency.

Several alternatives to traditional reinforced concrete structure and brick wall buildings have emerged, including the lightweight steel-framed (LSF) buildings. Given its advantages (economical, functional, environmental, etc.), the market share of LSF construction system grew significantly, mainly in low-rise residential buildings, making this kind of construction more attractive and popular [3]. Some of these advantages are as follows: high architectural adaptability [4, 5]; reduced weight; cost-efficiency [6]; exceptionally solid relative to weight; rapid on-site erection; excellent stability of shape in case of humidity; easy to prefabricate; and great potential for recycling and reuse, increasing building sustainability [1, 3, 7]. Section 2 of this chapter presents a brief overview of the LSF construction system, which includes the materials used, its classification regarding the position of thermal insulation, and the methods for manufacturing and framing.

Regarding sustainability, to perform a life cycle analysis of a building is essential to quantify both embodied and operational energies. To increase the sustainability label, it is vital to reduce both types of energies. This chapter focuses on the operational energy related to thermal behaviour improvement in LSF elements or components and energy efficiency of LSF buildings.

Some advantages of LSF construction system have been mentioned. However, when not correctly addressed during design stage, the LSF construction system may have also some drawbacks which could penalise its thermal behaviour and energy efficiency. Thermal bridges (TB), originated by the steel studs and the reduced thermal inertia (TI), are two major examples of these possible drawbacks. These issues related to the thermal behaviour of LSF elements are further detailed in Section 3.

Since the assessment of the energy efficiency of LSF buildings depends on so many factors and should be made in a holistic manner, this is not straightforward [1]. The parameters with influence on thermal performance and energy efficiency of LSF buildings could be grouped into four key factors [1]: climate [8, 9]; building envelope; occupants behaviour; and buildings systems. In Section 4 of this chapter, each one of these key factors will be further analysed.

Several tools to evaluate the energy and environmental performance of buildings in steel have been implemented. One example is the SB_Tool, also designated as ESSAT (early stage sustainability assessment tool), and developed by SB_Steel research project partners, mainly by the University of Coimbra research team, for the evaluation of the life cycle environmental performance of a building, which is freely available online [10]. This tool was an outcome of the European research project "*SB_Steel—Sustainable buildings in steel*" [11–13].

Some case studies related to the thermal behaviour of LSF elements and the energy efficiency of LSF buildings are briefly presented in Section 5 of this chapter.

http://dx.doi.org/10.5772/66136

2. Overview of LSF construction system

This section provides a brief description of the lightweight steel-framed (LSF) construction system. First, an overview of the main materials used in this construction system (structural cold-formed steel sections, sheathing panels and insulation materials) is presented. It continues with the typical classification of LSF construction components, concerning that the thermal insulation location within these components is described and concludes with a concise overview of the manufacturing processes and the framing methods.

2.1. Materials

The LSF dry construction system typically makes use of the following three main types of materials [1]: (i) structural cold-formed steel sections; (ii) sheathing panels (e.g. gypsum plasterboard and OSB—oriented strand boards; and (iii) insulation materials (e.g. expanded polystyrene for ETICS—external thermal insulation coating system—and mineral wool used within the walls and slabs). There are also some complementary additional materials like self-drilling screws for joining and fastening, air tightness and waterproof membranes, and of course the finishing cover layer. **Figure 1** illustrates a low-rise LSF residential building under construction, namely the cold-formed steel structure frame (**Figure 1a**) and after the setting up of OSB sheathing layer (**Figure 1b**). Notice that, as usual, to avoid ground humidity related problems, there is an elevated reinforced concrete ground floor [14].

(a)

(b)

Figure 1. Example of a low-rise LSF residential building at construction stage [14]. (a) Steel frame; (b) OSB external layer.

2.1.1. Cold-formed steel profiles

There are several cold-formed cross-sectional steel profiles, most of them identified by a letter (e.g. U, C, Z). The structural and functional performance depends on this cross-sectional shape, existing some special profiles with increased thermal (e.g. slotted web profiles) and acoustic performance (e.g. resilient profiles). To avoid corrosion and to increase durability, the steel studs are usually galvanised. The galvanisation process is often the hot-dip zinc immersion technique [3]. These steel studs are used in all LSF building components, namely external and internal walls, roofs and slabs, except ground floor slab, which is usually in reinforced concrete, as previously mentioned and illustrated in **Figure 1b**.

2.1.2. *Sheathing panels*

OSB and gypsum plasterboards are the most standard sheathing panels for the outer and inner layers of LSF construction elements (e.g. walls), respectively (**Figure 2**). Notice that, besides their covering function, these panels may have also a relevant structural role in load-bearing walls regarding horizontal loads, for example, wind [15]. Besides walls, OSB panels could also be used in slabs (e.g. floors and roofs), its thickness being usually greater than in walls. Furthermore, to increase thermal inertia/mass and reduce floor vibrations, the use of a top thinner concrete/mortar layer (e.g. 50 mm) could be advantageous [1].

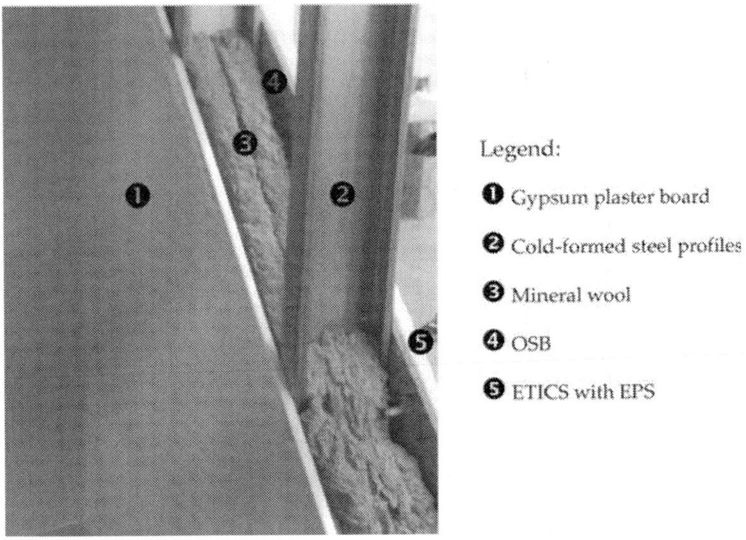

Figure 2. Materials in a LSF wall crosssection [30]. Legend: Gypsum plaster board; Cold-formed steel profiles; Mineral wool; OSB; and ETICS with EPS.

2.1.3. *Joining and fastening*

There are several methods for joining and fastening construction elements (e.g. two steel profiles or panels to LSF structure), being this issue very relevant for the speed of erection and for the mechanical resistance of the assembled structure. The use of self-drilling screws is the most usual fastening method, given its advantages, for example, stronger connection and higher durability, when compared with the use of nails [3]. Self-drilling screws are usually fabricated from heat-treated carbon steel or from stainless steel. There are several thread types for thread-forming screws, including for fastening thin sheets to thin sheets and for fixing to steel bases of greater thicknesses (greater than 2 mm or up 4 mm) [1].

2.1.4. *Thermal insulation materials*

As mentioned before, mineral wool is very often used between the steel sections as thermal and acoustic insulator (**Figure 2**). Besides, this insulation material is incombustible providing an improved fire resistance to LSF components. The use of expanded (EPS) or extruded polystyrene (XPS) is also very usual in the ETICS given its suitability to reduce thermal bridging-originated by the steel frames since, unlike the mineral wool batt insulation, it is a continuous thermal insulation layer (**Figure 2**) [16].

2.1.5. Wind and air tightness membranes

The adequate use of wind and air tightness membranes is very relevant to control heat losses due to air infiltrations in LSF buildings, mainly in cold climates [1]. In order to make sure that these air tightness membranes are correctly installed and the air infiltration rate is reduced, a "blower door test" or fan pressurisation method should be performed [1]. Besides heating energy reduction, another advantage of the adequate use of these membranes is the mitigation of the risk for interstitial condensation, given the resulting reduction in the moisture content inside the LSF element [3].

2.1.6. Finishing options

The most usual finishing coating layers are ETICS and gypsum plasterboards for outer and inner sides of exterior walls, respectively, being the gypsum plasterboards also very common in ceilings. However, the LSF construction may have any finishing covering layer as a traditional building with reinforced concrete structure and ceramic brick walls [3].

2.2. Classification of LSF construction

Usually, depending on the position of thermal insulation materials, the LSF construction elements are classified as cold, hybrid and warm frame construction [7], as illustrated in **Figure 3**. When all the thermal insulation is placed between steel studs (batt insulation), it is called "cold frame construction" (**Figure 3a**), since there is higher heat loss across the steel thermal bridge and consequently the steel temperature decreases leading to a higher risk of interstitial condensation, which could be particularly relevant in colder climates. The most usual LSF construction type is the hybrid one (**Figure 3b**) where, besides the batt insulation, there is also a continuous layer of thermal insulation, usually in the outer side (ETICS). In cases when all the thermal insulation is placed outside the steel framing, the steel frame is warmer (compare **Figure 3c** with the other two **Figure 3a, b**) and therefore, it is called "warm frame construction". Regarding the characteristics of thermal-hygrometric behaviour, the best option is warm frame construction, given the continuous thermal insulation and consequent lower thermal transmission value (U), reducing the risk of interstitial condensation [1]. However, in this option, the walls are thicker, and therefore, the net floor area could be diminished.

2.3. Manufacturing and framing methods

The main LSF construction framing methods are [1]: (i) stick-framing (or stick-built); (ii) panelised (or areal, "2D"); and (iii) modular (or volumetric, "3D"). Stick-framing was the first framing method to be used, where the steel studs are assembled together on-site, increasing flexibility and reducing planning needs. Given the great suitability of LSF construction for industrial modular prefabrication and consequent higher erection speed and improved quality control, the panelised and the volumetric system is being used more often. In these framing methods, the wall panels, floor cassettes and the 3D modules are prefabricated in factory with suitable dimensions to be transported to the construction site, where they will be assembled.

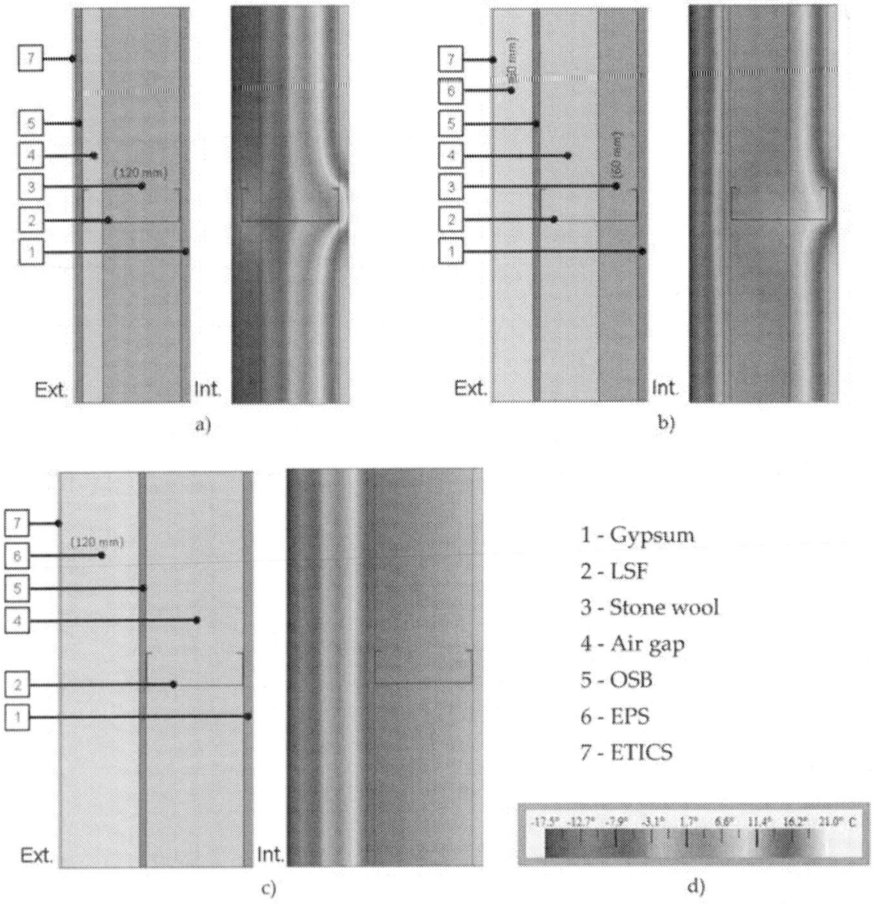

Figure 3. Classification of LSF walls, inside temperature distribution and thermal transmittance values (U). (a) Cold frame construction; U = 0.5255 W/m²/K. (b) Hybrid construction; U = 0.3856 W/m²/K. (c) Warm frame construction; U = 0.2828 W/m²/K. 1—Gypsum. 2—LSF. 3—Stone wool. 4—Air gap. 5—OSB. 6—EPS. 7—ETICS. (d) Materials and colour legend.

In order to take advantages of both 2D panel and 3D modular LSF construction, the "Hybrid" modular and panel could be also used as detailed for a case study building in UK by Lawson and Ogden [17]. Moreover, to extend the use of LSF construction to taller multi-storey buildings, it is possible to make use of an additional primary steel frame in order to provide the adequate structural stability to the building [17].

http://dx.doi.org/10.5772/66136

3. Thermal behaviour of LSF elements

This section includes a brief description of the thermal behaviour of LSF elements. It is easy to design LSF building envelope elements (e.g. walls, floors and roofs) with high thermal resistance values, even using lower thicknesses, while saving net construction areas. This goal is achieved by using thermal insulation materials [1]. Given the specificities of LSF construction [7], special attention should be given to the design stage in order to mitigate thermal bridges (originated by steel studs) and increase thermal inertia (if needed). These two key issues are addressed next.

3.1. Thermal bridges

Given the high thermal conductivity of steel, the design of building envelope components should follow certain rules in order to minimize the effects of thermal bridges (TB). Some examples of these design rules are [1]: if possible, avoid any interruption of the insulating layer; at least one third of the thermal insulation should be continuous (preferably external insulation as mentioned before in Section 2.2); at junctions of building elements, the insulating layers have to join at full width; if interrupting the insulating layer is unavoidable, use a material with the lowest possible thermal conductivity; keep façade geometry simple; and openings (windows and doors) should be installed in contact (at least partially) with the insulation layer.

Moreover, there are some specific parameters with direct influence in the thermal transmission of LSF construction elements including: the crosssection and number of steel frames; the thickness of the steel; the spacing of the steel studs; and the length of the web and flanges. Furthermore, there are several additional measures to mitigate the TB effects as illustrated in **Figure 4**. Since the major heat losses may occur across the steel frames, the use of thermal break strips along the studs (**Figure 4a**) is a possible strategy. The efficiency of this TB mitigation measure will increase with the use of high-performance thermal insulation strips (e.g. aerogel). Another approach could be the use of slotted steel studs as illustrated in **Figure 4b**. This strategy will increase the thermal performance of LSF elements (lower U-value) but will also decrease its mechanical resistance, which should also be taken into account for load-bearing studs [15]. The third example presented in **Figure 4** is the use of flange stud indentation. The geometry of the flange reduces the contact area between the sheathing panels originating a sort of thermal break given the small air gap created. The increase in the flange indentation size will also improve the thermal performance as illustrated in **Figure 4c**. In this case, the wall thermal resistance improvements were 9 and 16%, having as reference a standard steel stud.

Thermal bridges may have a very important influence on the energy efficiency of buildings, particularly in cold climates regarding the energy for space heating [1]. This issue is even more relevant in LSF buildings given the high thermal conductivity of steel [16]. Therefore, special attention should be given to thermal bridges mitigation at design stage, and as exemplified here, there are today several strategies available.

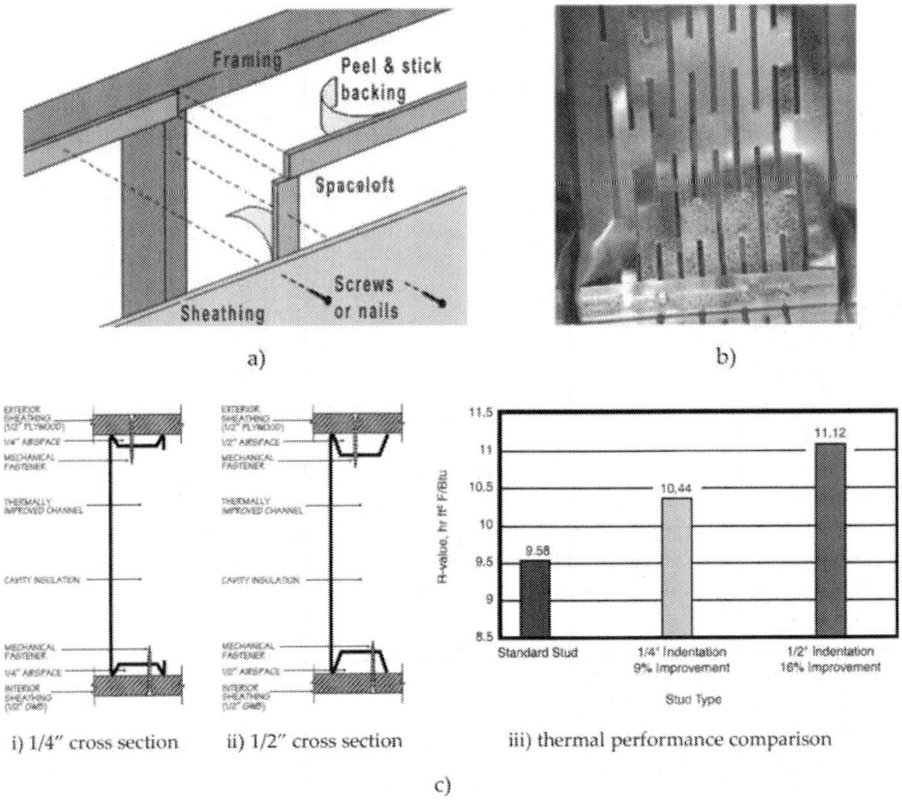

Figure 4. Strategies to mitigate thermal bridges in LSF construction elements. (a) Thermal break strips [18]. (b) Slotted steel stud [19]. (c) Flange stud indentation [20]: (i) 1/4" cross section. (ii) 1/2" cross section. (iii) thermal performance comparison.

3.2. Thermal inertia

LSF buildings exhibit lower thermal inertia (TI) when compared with traditional buildings with reinforced concrete structure and ceramic brick walls, given its reduced weight and consequent minor thermal mass. In practice, this means that LSF buildings may have higher internal temperature fluctuations. Therefore, it is important at design stage to adequately delineate the dimensions, exposure and shading strategies of glazed openings, with the aim to control solar heat gains, mainly during cooling season to prevent overheating.

It should be noted that a higher TI in buildings is not always advantageous regarding energy efficiency. Whenever the building has an intermittent occupation, as happens in many of the residential buildings during weekdays, this apparent drawback could be an advantage! In conventional low thermal mass LSF buildings, when the airconditioning system is turned on it will be much more easy and quick to cool/heat the building and achieve the required comfort temperature, thus reducing energy consumption and increasing energy efficiency.

However, if we are to take advantage of passive solar heating, a "mechanism" that stores solar thermal energy during the day and releases it during the night will be required. In this case, the thermal mass inside the building would be very useful. Therefore, sometimes in this circumstance, it is convenient to increase the thermal mass inside buildings, that is, its TI. **Figure 5** illustrates several strategies to increase TI inside LSF buildings. The use of ETICS (**Figure 5a**), that is, external thermal insulation allows not only to increase TI but also to mitigate TB, since it is a continuous thermal insulation layer. The second example illustrated in **Figure 5** is the use of massive materials (e.g. stones) in order to absorb and store heat. In this example, the stone wall was placed in front of a window to easily capture the solar heat, similarly to an internal Trombe wall. **Figure 5c** displays the average outside air and ground monthly temperatures (2 m deep) for Coimbra (PT) [21], as well as the difference between both temperatures. This temperature difference is not constant and is more significant during winter and summer time, reaching a value of +7.8°C in December and -7.0°C in July. Notice that, the ground is cooler during the cooling season and warmer during the heating season, that is, favourable in both seasons. There are several ways to take advantage of this air-ground temperature difference. The use of a ground-source heat exchange (GSHE) system based on air [22] or liquid (e.g. glycol fluid) flow through buried pipes is a possible strategy.

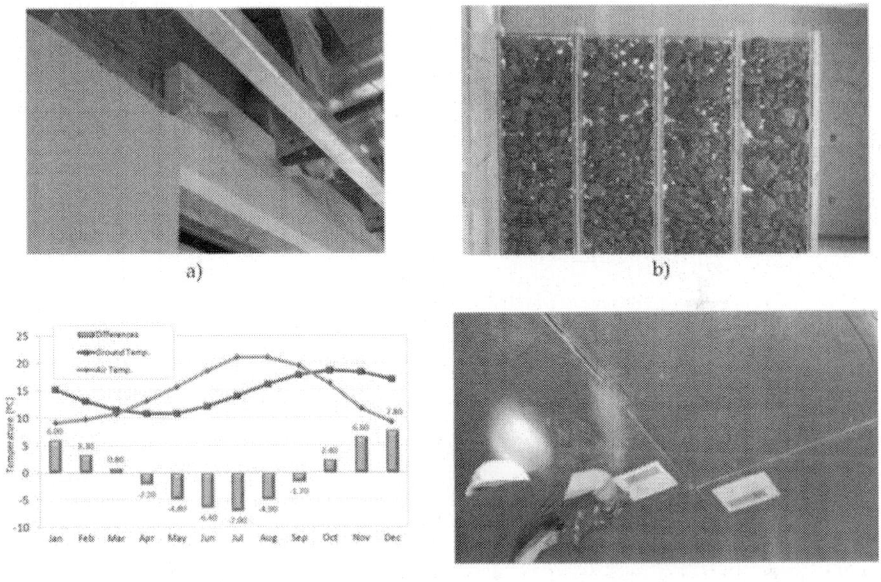

Figure 5. Some strategies to increase thermal inertia of LSF buildings. (a) External thermal insulation. (b) Use of massive construction materials [31]. (c) Make use of the enormous ground thermal mass. (d) Use of PCMs [23].

To conclude this set of examples, **Figure 5d** illustrates the use of a phase change material (PCM) in a ceiling (aluminium laminated PCM panel [23]). PCMs are able to store and release an enormous amount of heat whenever there is a temperature change that originates a phase change (melting or solidifying) given the so-called latent heat [24]. This latent heat allows the material to absorb or release heat without raising the material temperature, thus increasing the thermal inertia of the surrounding compartment.

Nowadays, a wide range of building materials or components containing PCMs can be found in the market [1]: boards for dry wall construction; plasters (e.g. gypsum, cement, clay); suspended ceiling tiles; internal window louvres; heat storage tanks; and under-floor heating system, etc.

Given the usual lower thermal mass in LSF buildings, the performance of PCMs is enhanced in this type of construction. However, the efficiency of PCMs in buildings depends on a lot of factors. Some relevant aspects should be taken into account such as [25]: (i) location in the building; (ii) their volume and thermophysical properties; (iii) the phase change temperature range; (iv) the latent heat capacity; (v) the climatic conditions; (vi) internal and solar heat gains; (vii) reflectivity and orientation of the surfaces; (viii) ventilation rates; (ix) HVAC controls; and (x) architectural characteristics. A case study will be briefly presented in Section 5.2 regarding the space heating/cooling energy performance optimization resulting from the incorporation of PCM drywalls in LSF residential buildings for different climates.

3.3. Energy efficiency of LSF buildings

Thermal behaviour and energy efficiency of buildings depend on a lot of factors. Moreover, its assessment should be performed in a holistic way, making its accurate evaluation/prediction very challenging. These parameters could be grouped into a set of four main key factors as illustrated in **Figure 6**: (i) climate; (ii) building envelope; (iii) building services; and (iv) human factors. These factors will be briefly described in the next sections.

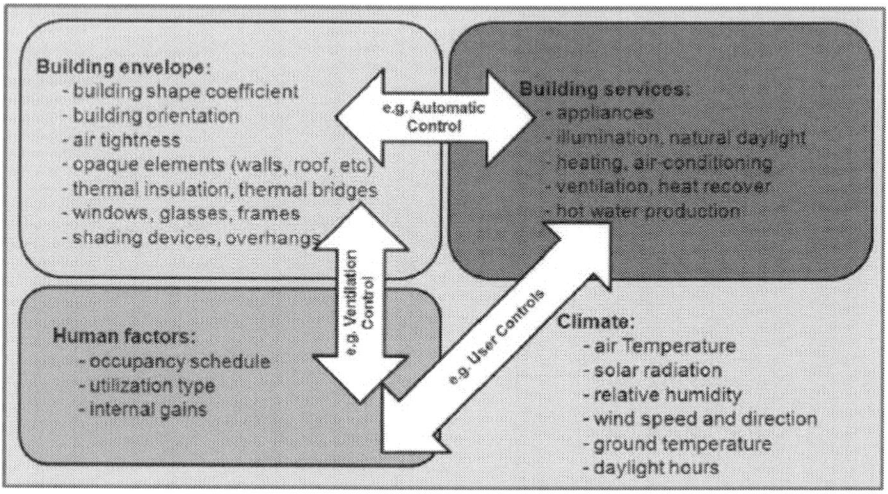

Figure 6. Key factors with influence on buildings energy consumption [1].

3.4. Climate

Climate is an external key factor with an impact on thermal behaviour and energy efficiency of buildings, mainly regarding energy for space heating and cooling. Obviously, climate depends on the building location. The main climate parameters are as follows: air temperature; solar radiation; relative humidity; wind speed and direction; ground temperature; and daylight hours.

The Köppen-Geiger climate classification [26] is one of the most widely used. In this climate classification, each climate is identified by a set of three letters. The first one represents the main climate classification: A—equatorial; B—arid; C—warm temperate; D—snow; and E—polar. The second set identifies the usual amount of precipitation: W—desert; S—steppe; f—fully humid; s—summer dry; w—winter dry; and m—monsoonal. Third one categorises the temperature: h—hot arid; k—cold summer; a—hot summer; b—warm summer; c—cool summer; d—extremely continental; F—polar frost; and T—polar tundra.

A common approach to characterise climate and relate outside temperature with the energy predictions for heating/cooling purposes is to make use of heating and cooling degree-days (HDD and CDD, respectively) having as reference a base temperature, for example, 18°C. **Figure 7** illustrates the average annual heating and cooling degree-days computed for the most relevant five European Köppen-Geiger climatic regions. The Figure clearly shows colder climates typical of Central (Cfb and Dfb) and Nordic (Dfc) European countries, where heating energy needs are largely greater than cooling needs. For southern European countries (i.e. Csa and Csb climate regions), the HDD are still high than CDD, but with much lower values when compared with the previous climate regions. Several case studies about the impact of climate on thermal behaviour and energy efficiency of LSF Buildings will be presented in Section 5.2.

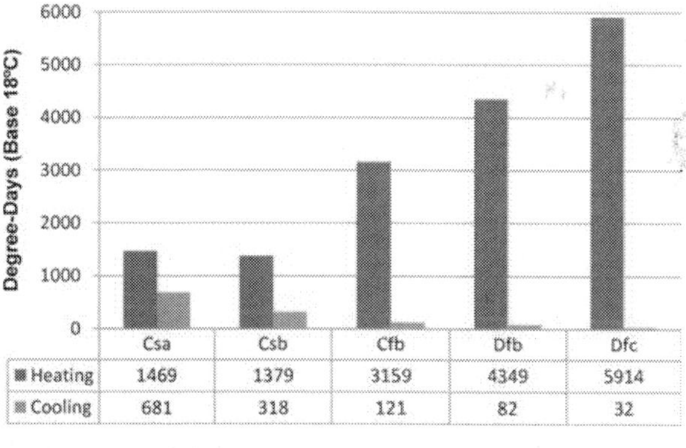

Figure 7. Average annual heating and cooling degree-days for five climatic regions [1].

3.5. Building envelope

The building envelope is another key factor to take into account the energy consumption of buildings [1]. As previously illustrated in **Figure 6**, some of the most important building envelope features are as follows: building shape coefficient; building orientation; air tightness;

characteristics of the opaque elements (e.g. walls, roof and floors) including thermal insulation and thermal bridges (see Section 3.1); thermal mass and thermal inertia (see Section 3.2); translucent elements (e.g. windows) including thermal and optical characteristics of glazing and frames; and shading devices, overhangs and sidefins.

The building envelope component responsible for the major heat losses during winter and solar heat gains during summer is usually the glazed openings (e.g. windows and doors). These undesirable heat transfer/gains could be mitigated by selecting glazing with lower thermal transmittance values and frames with thermal breaks, using insulated window shutters during night-time, designing adequate shading overhangs and sidefins, and suitable controllable shading devices (external ones are more efficient). Moreover, besides the thermal behaviour and energy performance of the building, the glazed building envelope is also very important for the thermal and visual comfort of building occupants as illustrated in **Figure 8**. In fact, the building indoor environment, for example, glare control, daylight and views, strongly depends on glazing features.

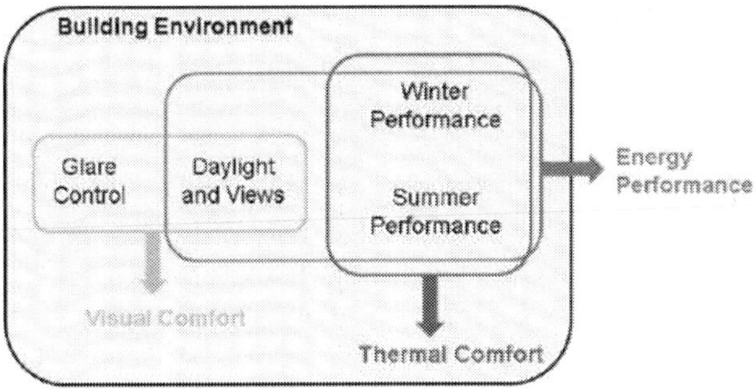

Figure 8. Glazing importance: energy performance and building environment [1].

Given the higher suitability of the LSF system for modular construction, Murtinho et al. [4] developed an architectural concept for multi-storey apartment building with LSF, which is illustrated in **Figure 9**. The main features of this design concept are modularity, easiness to build, energy efficiency and affordability, ensuring special flexibility, net area optimization and adaptability.

3.6. Occupants behaviour

The occupants behaviour is another very important issue regarding energy efficiency of buildings. In fact, buildings are inhabited and controlled by people who may contribute to increase or decrease energy consumption in the building. Some related examples are the occupation schedule (e.g. day, night or 24 h/day), the type of use (e.g. offices, residential, hospital) and the internal gains (e.g. the number of occupants, its metabolic activity level and equipment use). Obviously, the same building occupied by different people may have very different energy consumption values, given the differences in the occupants behaviour and comfort requirements regarding, for example, the heating and cooling air-conditioned temperature setpoints.

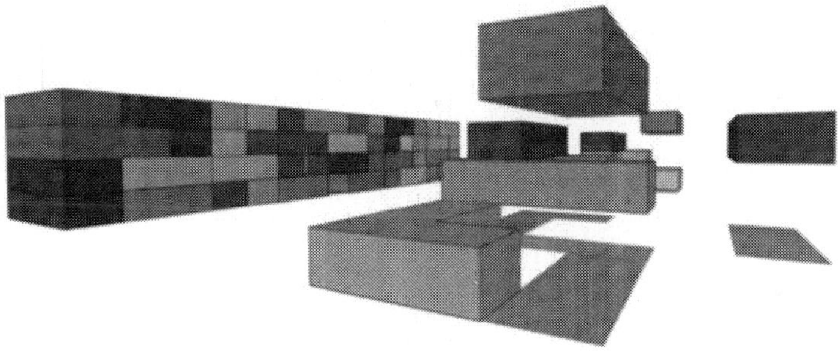

Figure 9. Modular architectural concept for multi-storey LSF buildings [4].

Offices have usually higher internal heat gains due to the intensive use of information technology equipment (e.g. computers and monitors) and consequent heat release. Moreover, offices are usually occupied during daytime, when external temperatures are higher. These two office features may lead to a higher cooling energy need when compared with other building typologies. A good example related to the metabolic activity of occupants is a gymnasium. In this case, the heat and moisture released by occupants could be very high due to the high metabolic activity and perhaps given the higher people density. Thus, cooling energy could increase and ventilation should be reinforced, not only to remove the air moisture but also the released metabolic CO_2. Moreover, these occupants may need a lower setpoint temperature to feel thermally comfortable, and this is an additional reason why energy for space cooling could be higher in gymnasiums.

3.7. Building systems

Another relevant energy efficiency key factor is the building systems. Some examples are as follows: illumination (control and efficient lamps); appliances; space heating and cooling; mechanical ventilation; hot water production; and mechanical ventilation heat recover. The control and efficiency of the equipment in use should be as good as possible in order to decrease energy consumption in the building. For instance, the electricity consumption of a thermal resistance heater (COP $\cong$ 1) when compared with an air conditioning system in heating mode (COP $\cong$ 4) will be about four times higher for the same amount of heat generated in the building. Moreover, the equipment systems should, whenever possible, make use of renewable energy sources. Two examples are solar collectors to produce domestic hot water and a biomass boiler for heating.

4. Case studies

In this section, several case studies related to thermal behaviour of LSF elements (e.g. walls) or components (e.g. earth to air heat exchanger—EAHE) will be briefly presented, namely the relevance of flanking thermal losses in LSF walls [27], the effectiveness of thermal bridges mitigation strategies [16] and the performance of an EAHE system located in the vicinity of a LSF building located in Coimbra, PT [22]. Furthermore, some additional case studies related to LSF buildings thermal behaviour and energy efficiency will be described, namely the "*Affordable Houses*" research project [5, 6], a parametric study regarding the thermal performance of LSF houses in Csb climatic regions [28], the impact of climate change on the energy efficiency of a LSF residential building [9] and the optimization of incorporation of PCMs in LSF houses in different climates [25].

4.1. LSF elements/components

4.1.1. Thermal bridges mitigation effectiveness assessment

In order to quantify and compare the effectiveness of several TB mitigation strategies in a LSF wall, Martins et al. [16] performed a parametric study using a 3D finite element method model previously validated against measured data [27]. **Figure 10** illustrates the studied reference wall model including the materials and the layer thicknesses (**Figure 10a**) and also the heat flux values predicted for the external surface of the LSF wall (**Figure 10b**). Several models were developed allowing the evaluation of the following TB mitigation strategies: (Model B) thermal break rubber strip; (Model C) vertical male or female studs; (Model D) slotted steel studs; and (Model E) fixing bolts instead of horizontal steel plate connection. The results (**Figure 11**) showed that the combination of all those TB mitigation strategies (Model G) leads to a reduction of 8.3% in the U-value, comparatively to the reference case (**Figure 10**), corresponding to 75% of the total impact of the steel thermal bridges.

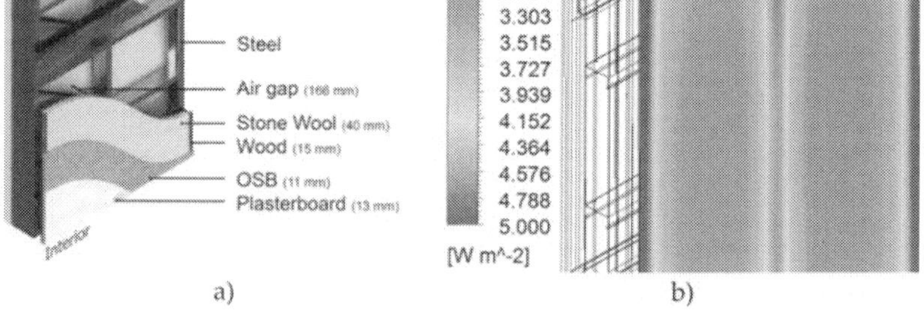

Figure 10. Reference LSF wall model used in the parametric study [16]. (a) Materials and thicknesses and (b) heat flux values on external surface.

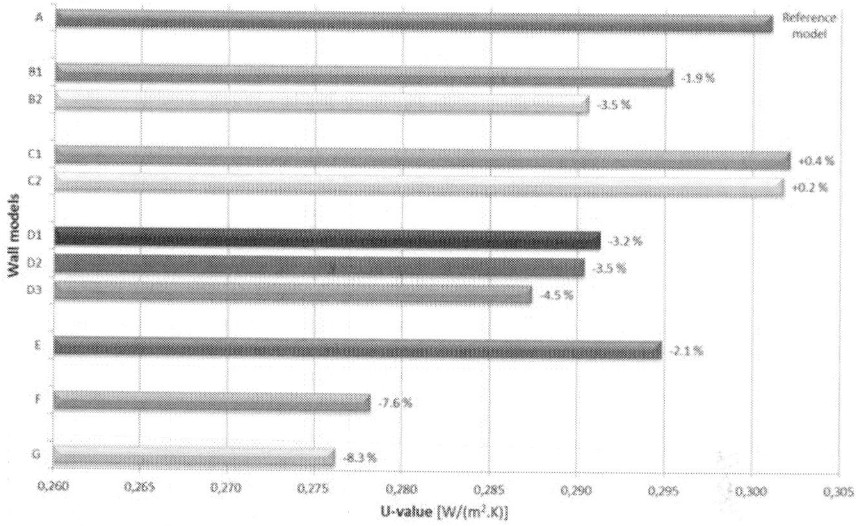

Figure 11. Results of the parametric study regarding the strategies for TB mitigation [16].

Additionally, making use of new insulation materials (aerogel and vacuum insulation panels) combined with the previously mentioned TB mitigation approaches, it was possible to significantly reduce the U-value of the wall (-68%), relatively to the reference case.

Martins et al. [16] also suggested some design rules for LSF elements: (i) at least 1/3 of thermal insulation should be continuous; (ii) the importance of the assessed single TB mitigation strategies is very reduced if the previous condition is verified; (iii) choose thermal profiles with higher number of narrow slots since they are more efficient; and (iv) use two layers of perpendicular steel profile studs avoiding trespassing the entire wall cross section with two parallel steel studs.

4.1.2. *Flanking thermal losses assessment*

Another issue instigated by the high thermal conductivity of steel is the increased importance of flanking thermal losses in the thermal performance of lightweight steel-framed walls. Santos et al. [27] performed an experimental evaluation of flanking thermal losses in a modular LSF wall tested in a steel gantry (**Figure 12**). Using an initial validated 3D detailed FEM model and also several others derived from this first model, they were able to evaluate the importance of several parameters in the flanking thermal losses, by computing the heat flux (**Figure 13**).

The most relevant parameters were, by decreasing order, the support steel gantry, the perimeter thermal insulation and the wall steel fixing elements. It was found that for a reference wall ($U = 0.30\ W/m^2/K$), the heat flux values changed from -22% (external surface) to +50% (internal surface) having as reference a wall with a flanking heat loss set to zero, that is, an adiabatic wall perimeter. Notice that, flanking heat losses are relevant not only in laboratory tests or numerical simulations but also in real buildings given the increased steel lateral heat exchange with the adjacent construction.

Figure 12. LSF wall tested in a climatic chamber [27]. (a) Inside view of the LSF wall structure. (b) External thermal insulation (EPS).

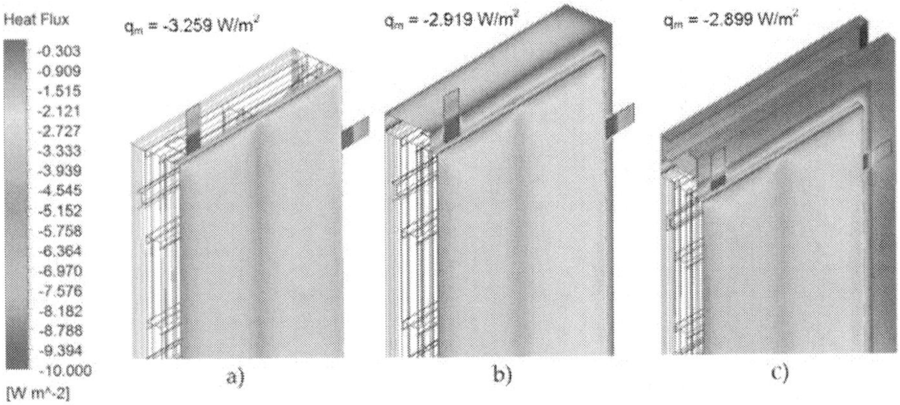

Figure 13. Heat flux predictions on the external surface of the wall for different 3D FEM models [27]. (a) Model B: "L" fixing elements. (b) Model C: "L"+XPS edge insulation. (c) Model D: "L"+XPS + Steel gantry.

4.1.3. EAHE system to increase thermal inertia

In addition to the building envelope (e.g. mitigating thermal bridges), the thermal behaviour and energy efficiency of LSF buildings could also be improved by making use of the huge thermal inertia of the ground. Santos et al. [22] monitored an earth to air heat exchanger (EAHE) system, located in Coimbra (PT), in order to assess its thermal and energy performance. This EAHE system consists of several buried ducts through which outdoor air for building ventilation is forced to flow by means of a fan. **Figure 14a** illustrates the buried pipes during the construction works of this EAHE, as well as the main relevant dimensions. Notice that, the fresh outdoor air is drawn into the EAHE through an inlet tower that contains a particle filter (**Figure 14b**).

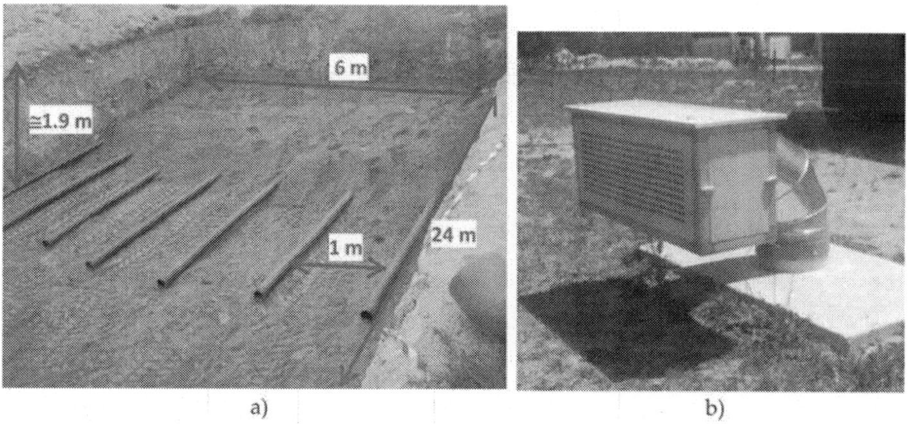

Figure 14. EAHE buried pipes and inlet filter. (a) EAHE buried pipes installation. (b) EAHE inlet filter.

Several parameters such as temperatures of the ground at different deeps, of the inlet and outlet air and the electric energy consumption of the fan have been recorded in different seasons of the year. **Figure 15** illustrates some of these recorded data and also the coefficient of performance (COP) of the EAHE system during November (heating season).

It was concluded that energy performance was higher in cooling mode (summer time), reaching an average COP of 1.7 during September, reaching a peak hourly value of 3.3. It was also observed that the control of the operation of these EAHEs is vital to optimize their energy efficiency, that is, the system should work only when it is useful to preheat (winter) or precool (summer) the air drawn into the building. Moreover, it was also found that the occupancy schedule of the building is another important parameter, that is, the system exhibits a higher heating performance during night-time (e.g. typical occupation schedule of residential buildings) and a higher cooling performance during daytime (e.g. typical occupation schedule of office buildings).

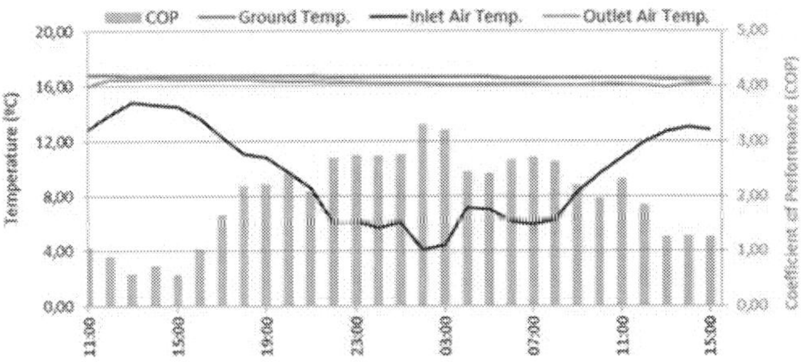

Figure 15. EAHE heating season performance: hourly values [22].

4.2. LSF buildings

4.2.1. "Affordable Houses" Portuguese proposal

The international research project *"Affordable Houses"*, involving eight countries (Brazil, Czech Republic, China, India, Poland, Portugal, Romania and Sweden), aimed to develop affordable and innovative housing concepts, which are culturally adapted to each country, using the LSF construction system. Moreover, each country proposal should be feasible, reproducible and exploitable. The total duration of this research project was 1 year, and it was divided into two stages: (1) pre-design stage and (2) design stage. The pre-design stage deliverables were as follows: (1) socio-economic evaluation; (2) traditional housing concept; (3) innovative concept; and (4) follow-up with general planning for 2nd stage. The design stage had two deliverables, namely (1) final design, including the detailed description of the technical solutions, and (2) socio-economic assessment.

The Portuguese proposal, prepared by a multidisciplinary team from the University of Coimbra, makes use of a modular LSF construction system developed by the national research team as illustrated in **Figure 16** and detailed by Murtinho et al. [5].

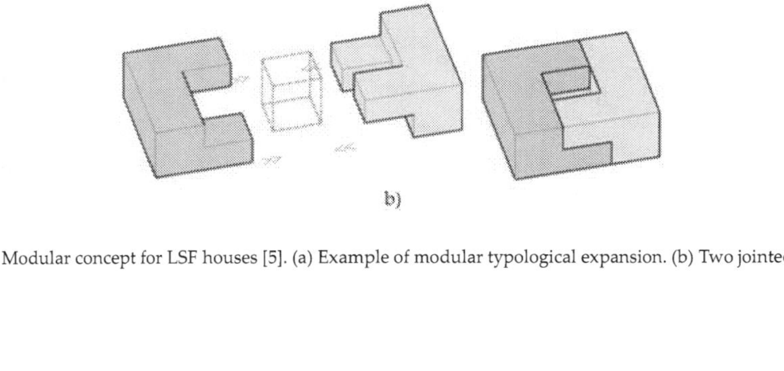

Figure 16. Modular concept for LSF houses [5]. (a) Example of modular typological expansion. (b) Two jointed houses.

The functional, structural and technological performance of the Portuguese proposal was evaluated and described by Santos et al. [6]. The building components and the functional requirements for proposed LSF residential building envelope were also presented, including energy performance, thermal and acoustic insulation, as well as the tools used in the design and performance assessment. The environmental performance of this house was also evaluated based on its carbon emissions. The thermal behaviour and energy efficiency of buildings were evaluated in accordance with the Portuguese regulation, also performing some advanced dynamic simulation using the *DesignBuilder* software as illustrated in **Figure 17**.

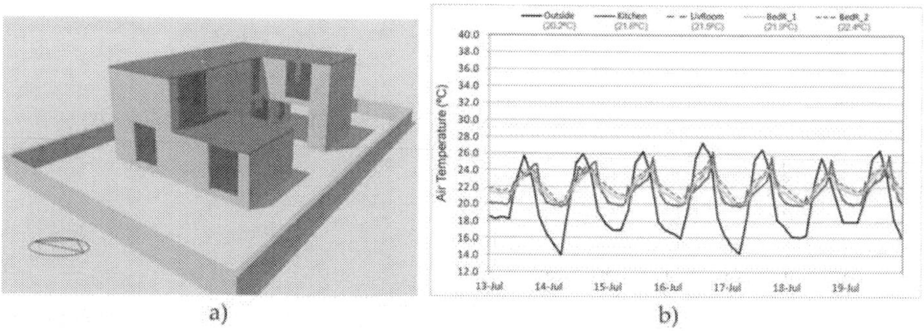

Figure 17. Modular Portuguese LSF house thermal performance assessment [6]. (a) *DesignBuilder* model. (b) Summer typical week.

4.2.2. Passive thermal performance in Csb climatic regions

Thermal performance of buildings could be assessed in an active mode (i.e. with the space cooling/heating equipment working) or in a passive mode (i.e. with the cooling/heating systems turned off). Santos et al. [28] performed a parametric analysis of the passive thermal performance of LSF residential buildings in Csb climatic regions located in southern European countries. With that purpose, a Portuguese low-rise residential building (**Figure 18**) was monitored in terms of its thermal behaviour, and an advanced dynamic *DesignBuilder* model was assembled (**Figure 18c, d**), calibrated and validated making use of the *in situ* recorded data. The relevance of several parameters (e.g. thermal insulation, ventilation, windows glazing, shading devices and overhangs) on the passive thermal behaviour of this building was evaluated making use of a previously validated model. Moreover, an optimum building envelope and operational control solution were specified, and design guidance was provided for the range of Csb climatic conditions. **Figure 19** illustrates how to use the suggested design guidance regarding two parameters: thermal insulation for roofs, walls and ground floor (**Figure 19a**) and overhangs ratio (**Figure 19b**) for the Genova (IT). The suggested simplified design process is very easy to use. Taking into account the average annual mean temperature for the building location, in this case 16°C, it is only needed to mark this value in the abscissa axis, intercept with the plotted line, and the recommended value is obtained in the ordinated axis.

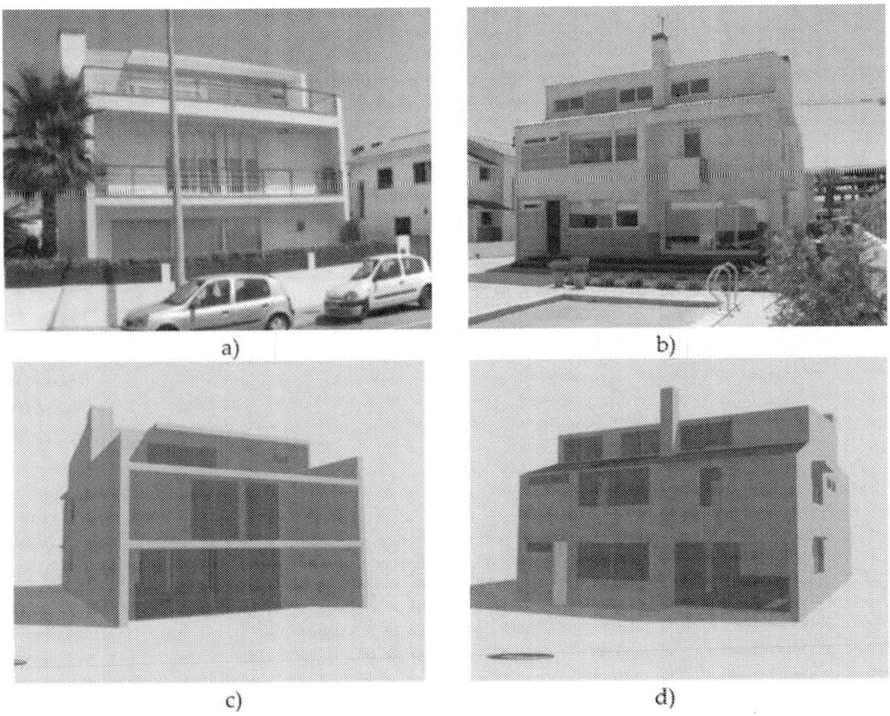

Figure 18. LSF residential building and *DesignBuilder* model [9]. (a) Front view. (b) Rear view. (c) Front view. (d) Rear view.

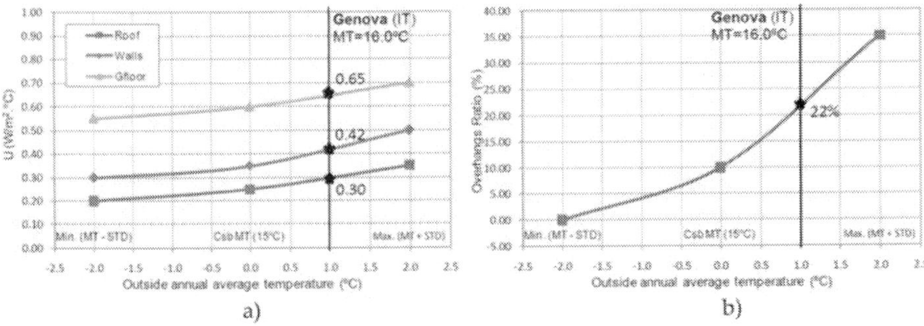

Figure 19. Design values suggested for Genova, Italy [1]. (a) Thermal insulation. (b) Overhangs ratio.

4.2.3. *Impact of global warming on the energy efficiency*

Obviously, global warming will induce changes on the thermal behaviour and energy efficiency of buildings. Santos et al. [9] assessed the impact of global warming on the energy efficiency of a LSF residential building (**Figure 18a, b**) based on the predictions of Intergovernmental Panel for Climate Change (IPCC) for southern European countries. With that purpose, an advanced dynamic simulation model developed in *DesignBuilder* software was calibrated against normative requirements regarding the thermal behaviour and energy consumption for space heating and cooling [29], and against a sophisticated computational fluid dynamics model (ANSYS CFX). Three climate scenarios were assessed, namely the annual recorded values for Coimbra city in Portugal (Scenario 1), and assuming an average temperature increase of +3°C (Scenario 2) and +6°C (Scenario 3). Besides climate, the energy consumption results for three building occupation schedule scenarios have been compared. Moreover, a set of winter and summer scenario combinations has been performed to predict the annual energy consumption and CO_2 emissions for a real occupation schedule scenario. **Figure 20** illustrates some of the results obtained. As expected, global warming will slightly reduce the energy for space heating but increase cooling energy. For the most probable climate change scenario predicted by IPCC (winter Scenario 2 and summer Scenario 3), an annual building energy consumption increase of 26.5% was projected. Regarding CO_2 production, the most likely increase in emissions was 15.0%.

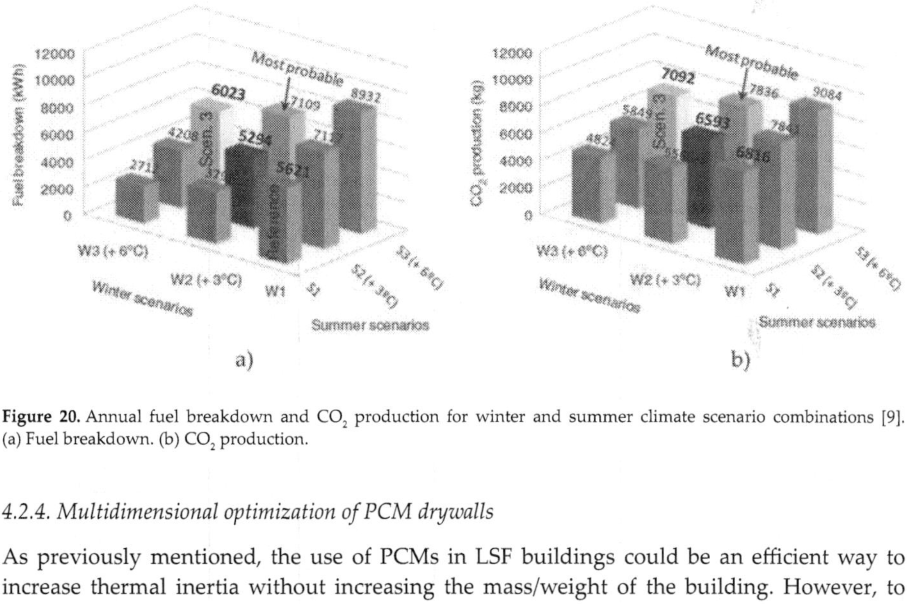

Figure 20. Annual fuel breakdown and CO_2 production for winter and summer climate scenario combinations [9]. (a) Fuel breakdown. (b) CO_2 production.

4.2.4. *Multidimensional optimization of PCM drywalls*

As previously mentioned, the use of PCMs in LSF buildings could be an efficient way to increase thermal inertia without increasing the mass/weight of the building. However, to optimize the efficiency of the PCMs in buildings is not an easy task since it depends on a lot of factors and they must be assessed in a holistic way [25].

Soares et al. [25] evaluated most of these factors by performing a multidimensional optimization of the incorporation of PCM drywalls in LSF residential buildings in different climates. This optimization was performed using *EnergyPlus* and *GenOpt* tools. **Figure 21** illustrates the model *EnergyPlus*, that is, a single-zone living room of a low-rise LSF residential building.

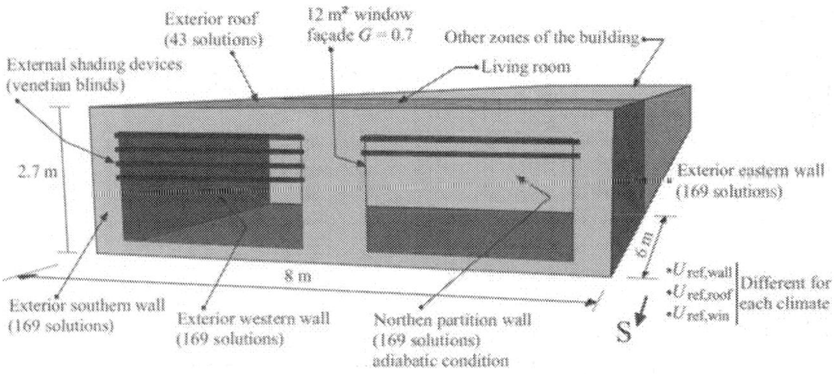

Figure 21. *EnergyPlus* model: single-zone living room of a low-rise LSF residential building [25].

The optimum solution for each climate (Csa-Seville, Csb-Coimbra, Cfa-Milan, Cfb-Paris, Dfa-Bucharest, Dfb-Warsaw, Dfc-Kiruna) was found considering a set of discrete variables in the model, namely the PCM enthalpy-temperature function, the PCM thermal conductivity-temperature variation, solar absorptance coefficient of the inner surfaces, the thickness and location of the PCM drywalls. To better simulate real-life conditions in the model, several parameters are included, mainly those related to the air conditioning setpoints, air infiltration rates, solar gains, internal gains from occupancy, equipment and lighting schedules.

It was concluded that the energy savings related to the use of PCMs in LSF construction were more evident in warmer climates. Given the higher daily external temperature amplitudes, PCM drywalls are particularly suitable for Mediterranean climates, with an expected energy efficiency gain of about 62% for Coimbra location (Csb climate). For the other climates/locations considered were obtained values between 10 and 46% regarding the energy efficiency improvement.

5. Conclusions

In this chapter, the thermal behaviour of LSF elements and energy efficiency of LSF buildings was presented, starting with an overview of LSF construction system including materials, classification, manufacturing and framing methods. The advantages of LSF construction were mentioned, and the two main potential drawbacks (steel originated thermal bridges and low thermal inertia) were addressed including several design rules to enhance the thermal behaviour of LSF elements. Moreover, the major key factors regarding the energy efficiency of LSF buildings were also assessed. Finally, case studies related to thermal and energy performance of LSF elements, components and buildings were presented.

LSF construction system has specific particularities (e.g. high thermal conductivity of steel and low thermal mass) that may have a relevant influence on thermal behaviour and energy efficiency of buildings. Therefore, special attention to design in terms of the mitigation of thermal

http://dx.doi.org/10.5772/66136

bridging and thermal inertia increase is essential to ensure a better thermal performance of LSF elements and an increased energy efficiency of LSF buildings. To illustrate this, several case studies were presented here, exemplifying the specificities of LSF construction system and its relevance in energy efficiency of buildings. Furthermore, in this design process, a holistic approach should be adopted in order to take into account the main energy efficiency key factors, that is, climate, building envelope, occupants behaviour (or human factors) and building systems. Only this way, it is possible to achieve "Energy Efficient Buildings", the title of this book.

Acknowledgements

The author would like to thank the ACIV ("Associação para o Desenvolvimento da Engenharia Civil") and the ISISE (Institute for Sustainability and Innovation in Structural Engineering) for their support.

Abbreviations and nomenclature

2D	two dimensional
3D	three dimensional
CDD	cooling degree-days
CO_2	dioxide carbon
COP	coefficient of performance
EAHE	earth to air heat exchanger
EPBD	energy performance building directive
EPS	expanded polystyrene
ESSAT	early stage sustainability assessment tool
ETICS	external thermal insulation coating system
EU	European Union
GHG	green-house gas
GSHE	ground-source heat exchange
HDD	heating degree-days
HVAC	heating ventilation and air conditioning
IPCC	Intergovernmental Panel for Climate Change
LSF	lightweight steel-framed
OSB	oriented strand board
PCMs	phase change materials
PT	Portugal
RES	renewable energy sources
TB	thermal bridges
TI	thermal inertia
U	thermal transmittance value [W/m^2/K]
UK	United Kingdom
XPS	extruded polystyrene

Author details

Paulo Santos

Address all correspondence to: pfsantos@dec.uc.pt

ISISE, Department of Civil Engineering, University of Coimbra, Coimbra, Portugal

References

[1] Santos P, Simões da Silva L, Ungureanu V. Energy efficiency of light-weight steel-framed buildings. In: 1st ed. European Convention for Constructional Steelwork (ECCS), Technical Committee 14—Sustainability & Eco-Efficiency of Steel Construction, ISBN 978-92-9147-105-8, p. 129; 2012.

[2] European Directive 2010/31/EU. Directive 2010/31/EU of the European Parliament and of the Council of 19 May 2010 on the energy performance of buildings (EPBD). Brussels, Belgium: Official Journal of the European Union; 2010.

[3] LSK. European lightweight steel-framed construction. Luxemburg: European Lightweight Steel-framed Association (LSK) and Arcelor; 2005.

[4] Murtinho V, Ferreira H, Antonio C, Simoes da Silva L, Gervasio H, Santos P. Architectural concept for multi-storey apartment building with light steel framing. Steel Constr. 2010;3(3):163–8.

[5] Murtinho V, Correia A, Ferreira H, da Silva LS, Gervasio H, Rebelo C, et al. Affordable houses: architectural concepts of a modular steel residential house. In: Cruz P, editor. Proceedings of the 2nd International Conference on Structures and Architecture. London: Taylor & Francis Group; 2010. p. 1325–31.

[6] Santos P, Mateus D, da Silva LS, Rebelo C, Gervasio H, Correia A, et al. Affordable houses (Part II): functional, structural and technological performance. In: Cruz P, editor. Proceedings of the 2nd International Conference on Structures and Architecture. London: Taylor & Francis Group; 2010. p. 1332–40.

[7] Santos P, Martins C, Simões da Silva L. Thermal performance of lightweight steel-framed construction systems. Metall Res Technol. 2014;111(6):329–38.

[8] Santos P, Gervásio H, Silva LS, Lopes AG. Global warming and the thermal performance of light steel residential buildings: parametric study. In: Topping BHV, Neves LFC, Barros RC, editors. Proceedings of the 12th International Conference on Civil, Structural and Environmental Engineering Computing. Stirlingshire: Civil-Comp Press; 2009. p. 1–17.

[9] Santos P, Gervásio H, da Silva LS, Lopesa. G. Influence of climate change on the energy efficiency of light-weight steel residential buildings. Civ Eng Environ Syst. 2011;28:325–52.

[10] SB_Tool. Website "SB_Steel Tool—Software to Evaluated Sustainability of Steel Buildings in Early Stages of Design": www.onesource.pt/sbsteel (last assessed: June 2016).

[11] Santos P, Gervásio H, da Silva LS. A simplified tool to evaluate the sustainability of buildings in steel in early stages of design. Matériaux Tech. 2016;104(103):1–12.

[12] Santos P, Martins R, Gervásio H, Silva LS. Assessment of building operational energy at early stages of design—a monthly quasi-steady-state approach. Energy Build. 2014;79:58–73.

[13] Gervásio H, Santos P, Martins R, Simões da Silva L. A macro-component approach for the assessment of building sustainability in early stages of design. Build Environ. 2014;72:256–70.

[14] Ferreira M. The construction system with lightweight steel structure (in Portuguese). Master Thesis in Civil Engineering, University of Coimbra; 2014.

[15] Veljkovic M, Johansson B. Light steel framing for residential buildings. Thin-Walled Struct. 2006;44(12):1272–9.

[16] Martins C, Santos P, da Silva LS. Lightweight steel-framed thermal bridges mitigation strategies: a parametric study. J Build Phys. 2016;39(4):342–72.

[17] Lawson RM, Ogden RG. Thin-walled structures "Hybrid"light steel panel and modular systems. Thin-Walled Struct. 2008;46:720–30.

[18] ThermaBlok. Website "ThermaBlok—Aerogel Insulation": http://thermablok.com (last assessed: June 2016).

[19] TUD. Website "ThermaChannel—Energy-Efficient Steel Framing": www.thermachannel.com (last assessed: June 2016).

[20] ThermaChannel. Website "ThermaChannel—Energy-Efficient Steel Framing": www.thermachannel.com (last assessed: June 2016).

[21] DOE-EnergyPlus. Website "DOE—EnergyPlus weather data base": www.energyplus.net/weather (last assessed: June 2016).

[22] Santos P, Sousa MA, Costa JJ. Modelling and performance analysis of an earth-to-air heat exchanger. In: Tadeu A, Ural D, Ural O, Abrantes V, editors. Proceedings of the 40th IAHS World Congress on Housing: Sustainable Housing Construction. Coimbra: ITeCons—Instituto de Investigação e Desenvolvimento Tecnológico em Ciências da Construção; 2014. p. 1–10.

[23] DuPont-Energain. Website "DuPont Energain—Thermal mass solution for drywalls": www.dupont.com (last assessed: December 2010).

[24] Soares N, Costa JJ, Gaspar AR, Santos P. Review of passive PCM latent heat thermal energy storage systems towards buildings' energy efficiency. Energy Build. 2013;59:82–103.

[25] Soares N, Gaspar AR, Santos P, Costa JJ. Multi-dimensional optimization of the incorporation of PCM-drywalls in lightweight steel-framed residential buildings in different climates. Energy Build. 2014;70:411–21.

[26] Köppen-Geiger. Website "Köppen-Geiger climate classification". http://koeppen-geiger.vu-wien.ac.at (last assessed: June 2016).

[27] Santos P, Martins C, da Silva LS, Braganca L. Thermal performance of lightweight steel framed wall: the importance of flanking thermal losses. J Build Phys. 2014;38(1):81–98.

[28] Santos P, Simoes da Silva L, Gervasio H, Gameiro Lopes A. Parametric analysis of the thermal performance of light steel residential buildings in Csb climatic regions. J Build Phys. 2011;35(1):7–53.

[29] EN 15265. Energy performance of buildings—Calculation of energy needs for space heating and cooling using dynamic methods—general criteria and validation procedures. CEN—European Committee for Standardization. Brussels; 2007.

[30] Perfisa-SA. Website "PF Perfisa—Fábrica de Perfis Metálicos SA". www.perfisa.net (last assessed: June 2016).

[31] TreeHugger. Website "TreeHugger—Making Sustainability Sexy": www.treehugger.com (last assessed: June 2016).

Chapter 4

Building Typology Comparison Between Courtyard and Atrium Buildings: A Study of Thermal Comfort and Energy Performance Factors in Different Climate Zones

Enes Yasa

Additional information is available at the end of the chapter

http://dx.doi.org/10.5772/66669

Abstract

The aim of the study in this chapter is to investigate performances shown by courtyard buildings, used widely both as microclimate regulators and as city-wide climate stabilizers especially in the hot-dry climate regions. Furthermore, this study examines atrium buildings having an increasing usage rate in recent years and the presence of comfort problems in particular which have not been resolved for different climate regions. Wind velocity measurements are performed in 36 different points determined in X and Y directions and 17 different points in the Z dimension on the outside of the courtyard considered in this study. In addition, both atrium building typology model and courtyard building typology model are obtained by taking the average courtyard dimensions seen in many regions; by covering open space courtyard section of the geometry with a transparent glass, atrium and courtyard typologies can be obtained. Furthermore, thermal comfort states and energy performances of these two different building typologies in interior courtyard and in building internal volumes for hot-dry, hot-humid and cold climate region conditions as well as the effect of solar radiation values exposing the building surfaces and solar movements during the day on the thermal performance on the building are analysed with CFD FloEFD and Star CCM+ software.

Keywords: thermal comfort, building energy performance, courtyard and atrium buildings, air velocity, wind tunnel experimental, numerical analysis, CFD, FloEFD, Star CCM+

1. Introduction

The history of using courtyards could be traced 5000 years back to Egypt. It is one of the architectural elements which are used continuously in different climates and civilisations for thousands of years [1, 2]. The usage of atrium can be traced back to the nineteenth century in

Europe after development of steel and glass technologies. Nowadays, highly glazed atriums are constructed for many different purposes in different climates [3–5]. Contemporary architecture mainly focuses on environmental aspects especially energy conscious forms of buildings. In this regard, recently, certain architectural studies have been conducted to understand the effects of these transitional spaces on buildings' energy consumption [6]. Two main types of transitional spaces, courtyard and atrium, are used widely all over the world.

Architects look for most appropriate design strategies in climatic responsive designs to integrate them with building designs in order to enhance their performance. Using both courtyard and atrium in different climates without considering their performance in different climatic conditions will cause energy-related concerns, which should be eliminated. It is important for designers to understand how different climates affect energy performance of courtyard and atrium buildings and which building type is more energy efficient in different climates.

There are many types of architectural zones which moderates the outdoor and indoor climatic conditions without mechanical control systems. These zones are called transitional spaces. They can be closed, such as atrium, or semi-closed, such as balcony and porch, or open, such as courtyard and patio [7].

1.1. Definition of atrium buildings

Atrium-building typology is defined as building of two or more storeys high open space that is enclosed with a roof and having vertical volumes surrounded with usable areas. Atriums are open public spaces where people engage in activities, e.g. to walk around, meet, talk, wait and rest. When the evaluations are made in terms of the systems used in atriums, it can be seen that there is no common solution agreed upon. Because of their spatial role, atrium-building typologies which form the centre of buildings create a buffer zone between the building and the external environment whilst providing vertical and horizontal circulation between the layers. Atrium-building typologies play an important role for various purposes in multi-storeyed commercial and institutional buildings; for instance serving as foyers, building entrances, exhibition hallways. When they are compared with traditional buildings, atrium-building typologies require excessive energy consumption in terms of heating and cooling as well as ventilation, since they contain more complex air phenomena (such as the greenhouse effect, buffer zone and air layering). In addition, despite this excessive energy consumption, failure to provide user comfort conditions for a long period of time in an atrium typology building is problematic. Meeting the required performance criteria for reducing energy consumption and to provide user comfort in atrium type buildings is possible by careful selection of the glass system which is used particularly in the atrium hall and by designing the building geometry according to climatic conditions.

When atrium-building typologies are compared to classical buildings, due to the fact that they contain complex climate events (sera effect, tampon area, air layering, etc.); they require extreme energy consumption for heating, cooling and for ventilation purposes. In spite of this, extreme energy consumption, the failure to obtain comfort conditions for users in typology building is seen as a significant problem, especially when the correct glass selection in the glass systems used in atrium hole along with the suitable geometry design compatible with topography and weather conditions could potentially solve the problems for reducing energy consumption and to obtain user comfort in atrium-type buildings.

Atrium-building typology is a plan type used widely in Europe and worldwide, due to a variety of reasons such as the ability to create innovative and prestigious locations, formation of social and comfortable environments along with maximized advantages due to utilization of sunlight for natural lighting and for warmth, as well as the presence of natural ventilation in these buildings [8].

Since these regions are designed for the actions having no continuity, it is thought to be warmer in the summer and colder in the winter compared to the internal environmental conditions. However, this approach has changed over time to longer-term actions e.g. eating, drinking, sitting and these places are conditioned according to these new developed actions. To meet the increasing user demands together with the rich utilization purposes needed, complex conditions, such as heating, cooling, ventilation, air stratification, ensuring indoor air quality, acoustic and environmental system controls, are experimented [9].

The energy used to obtain necessary comfort levels related to actions of atrium-building typology reaches very high levels. In today's conditions, where the effective use of energy is discussed, the inadequacy of comfort conditions even with high-energy consumption and with the extreme energy usage in atrium-type buildings are discussed as a serious problem.

1.2. Definition of courtyard buildings

1.2.1. Courtyard as a climate moderator

When other studies about courtyard buildings are considered, the recent studies frequently focus on investigating thermal performance within the courtyard. These studies can be classified as air movement in the courtyard, courtyard-building-sun-shadow relationship and thermal performances of courtyard buildings in different climate regions. In a study carried out on evaluating the total-energy performance of a courtyard selection, energy performance of a courtyard building with the same geometry and ratio and the centre atrium energy performance is investigated comparatively [10–12]. In another study, where the effect of building form and its type on climatic performance for various climate regions are studied and from this perspective, the evaluation of the courtyard option is conducted in terms of the environment. Ratti et al. performed numerical analysis studies for the conditions described above [13].

When the studies related to buildings with courtyards are examined, the studies which analyse thermal performance in courtyards are seen very frequently [10]. These studies can be classified as air movement inside the courtyards, the relationship between courtyard-building-shading and the thermal performances of courtyard buildings in different climates [8, 14]. In a study which examined the total energy performance of the courtyard, the energy performance of central atrium and the comparisons with a courtyard building having the same geometry and scale have been examined [15, 16]. In the study where the effect of building style and form has on the climactic performance and where the evaluation of the courtyard option is conducted environmentally, Carlo Ratti, Dana Raydan and Koen Steemers have conducted numerical analysis studies [17].

In another study by Mohsen, which investigated the thermal performance of a courtyard building, evaluated the effects of the geometrical and physical parameters of the courtyard on the solar heat radiation exposed on the front side of the courtyard structure [18]. Muhaisen and Gadi performed several studies on courtyard type and courtyard buildings. These studies mainly focused on the effect of the courtyard type and solar radiation gain as well as the sun-shading effect. The objective of these studies conducted in 2006 was to investigate how to provide adequate amount of solar radiation to obtain the necessary heat for the building in the winter and inner courtyard envelope and courtyard type work needed to reduce the energy necessary for the cooling in summer or to provide sufficient shadow region [19].

In another study, the sun-shadow performance of a courtyard was examined and a mathematical model was developed in order to calculate the shaded and sunny areas of the courtyard building which was designed with a circular geometry. This developed model investigates the interaction between the sun and the courtyard buildings in circular geometry having any ratio or dimension. Muhaisen and Gadi examined the shading performance of polygonal (such as pentagonal, hexagonal or octagonal) shape courtyard types in their study investigating the courtyard style and options [19].

Studies related to thermal performance of the courtyard option using CFD are commonly seen in recent literatures. In the study by Rajapaksha, Nagai and Okumiya, they have examined passive cooling potential in single storey courtyard buildings with dense, massive envelope in a warm, humid climate and they have put forward the criteria and suitability of single-storied, large mass courtyard buildings for passive cooling [20]. In a study by Rajapaksha et al., the passive cooling potential of a single-storeyed highly massive building in a warm-humid climate was examined. They tested the presence of the inner yard space to minimize the heating conditions through an increase of natural ventilation and its optimization [20, 21].

1.3. Thermal comfort behaviours of courtyard and atrium buildings

When the past studies about the courtyard typology and atrium-building typology are examined, it is observed that the energy efficiency and how much the typology effects in terms of climatic comfort as well as how they behave, the relationship and the differences among them are seen to be far from having a comprehensive strategy. Therefore, the aim of the study in this chapter is to investigate what is the relationship between the courtyard and atrium typologies with the most commonly used courtyard geometries according to climatic comfort requirements which are needed for different climate regions and also the meteorological differences as well as the type of behaviour they have in terms of energy performances and comfort conditions.

2. Methodology

2.1. CFD-numerical analysis process

In this study, Star CCM+ software is used for CFD (computational fluid dynamics) analysis. For the hardware resource, a supercomputer located in the Energy Systems Laboratory at Texas Engineering Experiment Station has been used for analysis of the total energy performances of both building options in terms of energy gain-loss in the building within the cooling period in summer (July 21) and the cooling period in winter (January 21) for both building typologies and for three different climate regions are evaluated. From the raw data obtained through CFD Star CCM+, total heat transfer amount and solar radiation gain tables of vertical and horizontal surfaces for all the buildings related to the period of 24 hours encompassing January 21 and July 21 are created for all three climate regions. Numerical values obtained from the tables are reported separately according to the total volume of both the total surface area of the building and the total volume of the building. CFD Star CCM+ software used in this study has also considered and calculated the effect of the shadow regions on the courtyard and building surfaces through sundials during the day for the purposes of heating and cooling requirements of the courtyard building.

2.2. Experimental wind tunnel process

Interior courtyard wind velocity measurements performed in the experimental period of this study is conducted in the wind tunnel in Physical Environmental Control located in the Istanbul Technical University- Faculty of Architecture. Interior courtyard wind velocity measurements are performed between courtyard ground point and a height of 2h. Flow speed measurements are performed with computer aid under the guidance of 'Streamline 3.03' software. In order to obtain velocity distribution both with and without a model belonging to the observation room, 'Streamline 3.03' CTA (Constant temperature) hot-wire type anemometer made by DANTEC company has been used.

A part of the statistical analysis and visual expression of the measurements have been accomplished with 'DANTEC's anemometer and a commercial software called 'ACQWIRE' which has been developed for controlling traversing systems. Therefore, in addition to the anemometer, a PC and a printer constitute the main parts of the hardware.

The units used in the setup are composed of 'Traversing system', one 57 B120 motor controller for manual control and 56 G 00 CTA 'interface' computer connector and two arms moving in one dimension carrying measurement terminals (DANTEC), as well as Turkish Q keyboard and Mouse, P11 and P15 type probes. Measurements are made compatible with PC via a software package called 'Streamline 3.03'. 'National Instrument' is developed by 'DANTEC' for hot wire anemometers. With this software; obtaining data, transforming electrical voltage variation obtained from the hot wire probe into velocity, as well as storing the measured data as files, calculating turbulence intensity and average speed data and evaluating them by transforming all of these into the plots are possible.

2.2.1. Properties of the wind tunnel

Physical Environment Control Wind Tunnel of Istanbul Technical University (I.T.U) Department of Architecture is a subsonic, open rotated, closed jet wind tunnel with Eiffel-type absorption. Tunnel entrance starts with 2.50 × 2.50 m × 0.30 m bell mouth. Subsequently, there are two flow formatting sections. For the connection between the flows formatting section to the observation chamber, there is an adapter module passing from 2.00 m × 2.00 m cross-section to the 1.05 m × 1.05 m cross-section. Collector section, where the air absorption takes place, is built by a sheet metal with 4 m^2 of cross-section, 3.40 m height and 2.00 mm of thickness. The air absorbed in this section reaches to the observation room with 1 m^2 cross-section by making the air parallel to the tunnel with the help of the bell mouth collector. In the end of the diffuser, total of 5.96 m long providing circular cross section with a diameter of r = 1.64 m of 1.05 × 1.05 m^2 cross-section is connected to fan pulley with 0.52 m diameter, which is connected to another fan pulley with a 0.52 m diameter. The fan utilized in the tunnel is an axial fan. The power of the motor rotating the fan is 1.5 kW and the rotation speed is 1450 rev/min (**Figure 1**).

Thirty-six different measurement points are determined in *X* and *Y* dimensions in the courtyard and measurement profiles with 34 measurement points are identified in the *Z* dimension.

According to the profile, a total of nine measurement points which are spaced with a distance of 0.5 cm are determined in the interval between 0 and 4 cm. due to the region where the first boxes of 4.00 cm are located. In the next section up to 10.00 cm, six measurement points with 1.00 of separation are identified. A total of 14 measurement points are identified with 0.5 cm separation again for the region between 10.00 and 17.00 cm. Finally, five measurement points between 17.00 and 22.00 cm are identified. Then the probe has been placed in the observation room of the wind tunnel (**Figures 2** and **3**).

Measurement points located parallel to the side face of the observation room of the wind tunnel are the 'A.B.C.D.E.F.' points. These points are also parallel to the wind direction. '1.2.3.4.5.6.' measurement points are the measurement points which are perpendicular to the wind direction (**Figure 3**).

Figure 1. General view of the wind tunnel in Istanbul Technical University-Department of Architecture-Physical Environmental Control Laboratory.

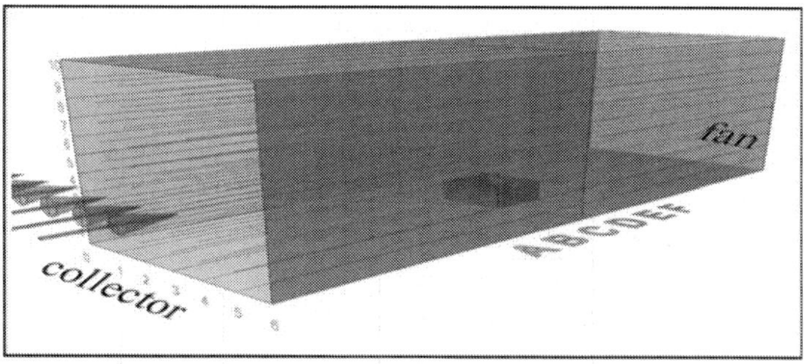

Figure 2. Model and measurement point axis views in the tunnel-positions of measurement axis on the lateral surface of the wind tunnel observation section.

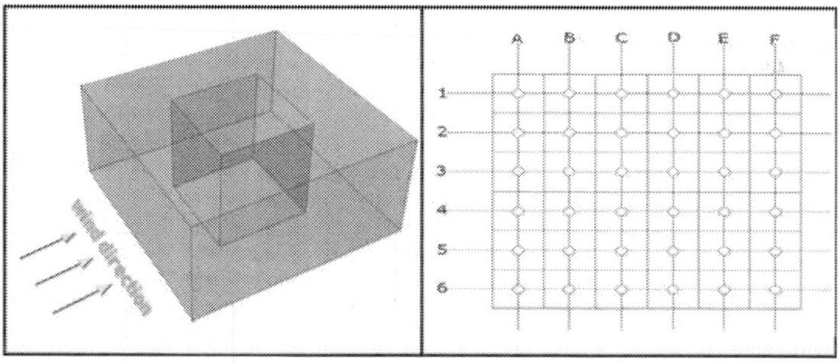

Figure 3. Thirty-six different measurement points for the model courtyard being measured.

Thirty-six different measurement points identified in the courtyard model used in this study are shown in **Figure 3**. There are total of 36 points including six of each in every dimension with 2.00 cm intervals in X and *Y* dimensions within 12.00 × 12.00 × 12.00 cm courtyard. These points are named as 'A1, A2, A3, A4, A5, A6—B1, B2, B3, B4, B5, B6—C1, C2, C3, C4, C5, C6—D1, D2, D3, D4, D5, D6—E1, E2, E3, E4, E5, E6—F1, F2, F3, F4, F5, F6' (**Figure 3**).

Interior courtyard measurement values of the discussed courtyard building typology and the measurement results made in the windward area are explained with table and plots. Then the measurement values in '0.00H–0.25H–0.50H–0.75H–1.00H–1.25H–1.50H–1.75H' levels are evaluated with tables and plots and these values are compared with each other (**Figure 3**).

2.3. Reference building considered for the courtyard and atrium typology

For both the chosen courtyard and atrium-type buildings, the numerical analysis and the courtyard building typology where the interior courtyard wind-velocity measurement are performed in the wind tunnel is designed as two floors, with 3.00 m floor height and with 14.00 × 14.00 × 6.00 m. outside of the building sizes and 6.00 × 6.00 × 6.00 m courtyard dimensions (**Figure 4**). At the same time, all geometrical information and building envelope information related with the courtyard typology is benefitted from the courtyard information considered in E. Yasa, V. Ok. 2014.

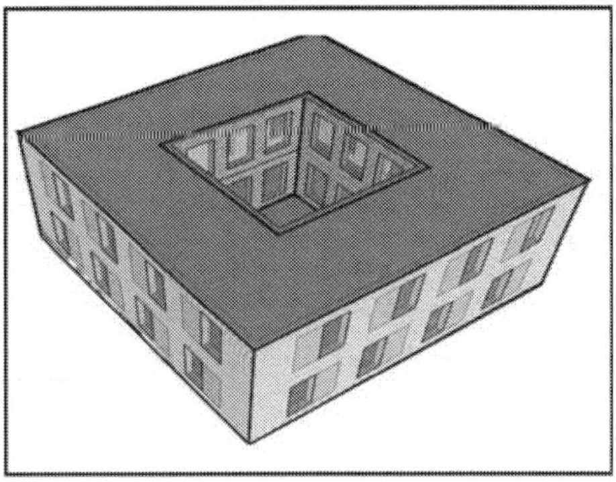

Figure 4. The plan and perspective for the reference courtyard-building (with two storeys, with a floor height of 3.00 m and with external building dimensions of 14.00 × 14.00 × 6.00 m and courtyard dimensions of 6.00 × 6.00 × 6.00 m.

3. Case study methodology and creating a model geometry for CFD

For the opacity and permeability values on the transparent surfaces of the building used for both types of buildings in this study; reflectivity is set at 10%, absorption is set at 26% and the permeability is set at 64%. Thickness, density, specific heat, thermal conductivity coefficient, solar radiation absorption, solar radiation reflectivity, surface roughness, layer number information as well as separate layers as mezzanine, roof tiles sitting on the ground and the layers on the tiling are identified separately (**Table 1**).

Materials	Thickness, dn (m)	Thermal conductivity calculation value, λh W/mk	dn/λn	Ud
Outer plaster	0.030	1.400	0.021	0.378
Rockwool	0.040	0.040	1.000	
Aerated concrete (with mortar proper to TS 4916)	0.200	0.140	1.429	
Inner plaster	0.020	0.870	0.023	
Total		**1/λ**	2.473	

Table 1. Information about the building envelope applied into the whole courtyard and atrium typologies.

These data about the building envelope are also listed in **Table 1**. The data used in the simulation program is completed by entering the values of volume ambient temperature, boundary condition of the surface and heat zones, absorbency of the surfaces, reflectivity, density, specific heat and thermal conductivity (**Table 1**). Data for Antalya which has a hot-humid climate character, Diyarbakır having hot-dry climate character and Erzurum with cold climate character are then entered into the Star CCM+ simulation software. In addition, structure envelope of the building, structure elements and data like reflectance, permeability of the building materials are also inputted. **Figure 5** shows the properties of the examined courtyard and atrium typologies the geometries.

http://dx.doi.org/10.5772/66669

Properties of the Considered Courtyard and Atrium Typologies	
Courtyard Typology	

Courtyard Dimensions	6.00m X 6.00m X 6.00m
Building Roof Area	160 m²
Building Exterior surface area(for Total Heat Transfer)	960 m²
Building Total Volume	960 m³
Courtyard Area	36 m²
Courtyard Volume	216 m³
Atrium Typology	
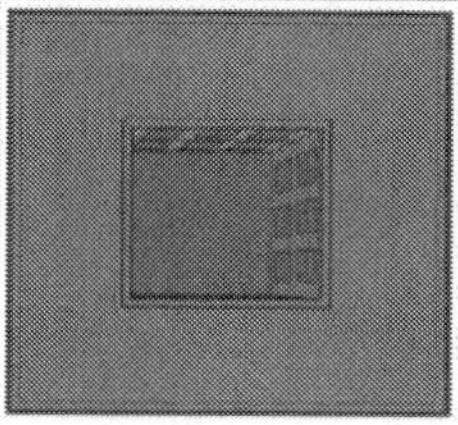	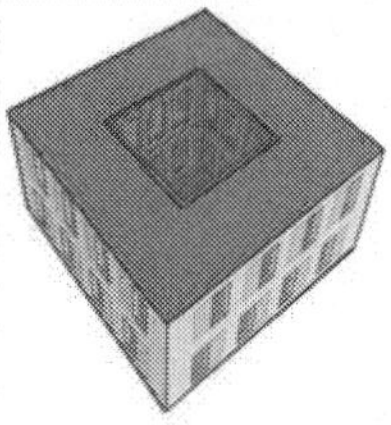
Atrium Dimensions	6.00 x 9.00 x 6.00
Building Roof Area	184 m²
Building Exterior surface area(for Total Heat Transfer)	1104 m²
Building Total Volume	1104 m³
Courtyard Area	54 m²
Courtyard Volume	324 m³

Figure 5. Properties of the considered courtyard and atrium typologies.

4. Limitations and assumptions in the case study

This study is limited to comparative analysis between courtyard and atrium typology options considered for application in plot centres of 'Hot-Dry Climate', 'Hot-Humid Climate' and 'Cold Climate' with different characteristics dominant in Turkey. The courtyard and atrium are of same size and volume. Total of analysis consists 24 hours of analysis for each option has been made separately on hourly basis. For each building typology, 7th month and 21st day, in other words, July 21 constituting the example of the hottest period has been elected as the average of the long-term meteorological data for summer months or cooling period instead of all months and days of a year; and 1st month and 21st day, in other words, January 21, the example of the coldest period, has been elected as the average of the longest period meteorological data average for the winter months or heating period. To represent the three distinct climatic regions examined in the study, 'Diyarbakır-DB' was chosen for the hot-dry climatic region, 'Antalya-ANT' for the hot-humid climatic region and 'Erzurum-ERZ' for cold climatic region and long-term average meteorological climate data pertaining to such provinces were used.

The reason why these days are discussed is that data where average values are frequently encountered have been found out for the heating and cooling periods having examined the long-term meteorological data. The limitation of the courtyard and atrium building transparent surface rates, however, the transparent surface ratio of the courtyard option in non-building facades is 20%, the ratio of opaque surface is 80% and such ratio is 40% for transparent surface ratio and 60% for opaque surface ratio in building courtyard facades. Opaque and transparent surface for atrium building is exactly the opposite to solar heat gain, all thermal factors were kept the same in the research for the purpose of defining the building's energy needs by estimating to what extent solar radiation is affected by various building typologies. Mechanical HVAC was not included among courtyard building options. The only cause of heat gain is expected to be solar heat gain. Every courtyard and atrium building was considered to have an average floor height of 3.00. In both courtyard and atrium building options, the building was considered to have a comfort limit temperature value of 25°C for heating and cooling loads inside. The user entered the meteorological and topographic data for the building's climatic region. The index of wind direction and intensity, outdoor weather temperatures, the region's sky cloudiness and direct and common solar radiation intensity were generated based on the data from Republic of Turkey General Directorate of Meteorology, which constituted the meteorological data entered to the software.

5. Boundary conditions

Naturally ventilated buildings are subjected to external climatic conditions in terms of their architectural design. The interaction between the indoor and outdoor environment results in an internal airflow pattern. The indoor airflow pattern is significantly affected by natural ventilation and outdoor conditions which in turn directs occupants' thermal sensations. The solar radiation within the building produces heat gain which is then thought to cause buoyancy-driven natural ventilation. This natural ventilation is believed to direct the building's natural ventilation.

Defined as the product of the density, thermal conductivity and specific heat capacity, the heat flux coefficient quantifies a material's ability to absorb heat. It was found that the heat flux coefficient reflects the influence on thermal comfort of different surfaces. Therefore it is used as the basic thermo-physical property which defines materials. Various layers of each element of the building (walls, roof, floor) are considered to have heat flux coefficients as design variables. Various layers' thicknesses in each building element are also considered as design variables [23].

Taking into account the courtyard and atrium building, the data is intensity: 2290 kg/m^3, specific heat capacity C_p: 840 J/kg K, thermal conductivity: 0.96 W/m K, thickness: 2.00 cm for transparent surfaces and thickness: 115 cm, intensity: 1590 kg/m^3, specific heat capacity (C_p): 850 J/kg K, thermal conductivity: 0.65 W/m K for opaque surfaces. k-€ (standard model); near wall treatment: standard wall functions were used as turbulence model, while solar calculation was used for insulation calculation.

Courtyard and atrium options can include materials with different material characteristics, such as air as building shell and fluid and as building components, floorings, doors, windows, walls and roofs. Intensity: 1.2256 kg/m^3, specific heat C_p: 1006.43 J/kg K, heat conduction: 0.0242 W/m K and viscosity: 1.7894e-05 kg/m s are respective values accepted for fluid air. Transmission coefficient was also taken into account; 0.3499 W/m K for doors, 1.3944 W/m K for walls, 0.919 W/m K for roofs and 1.05 W/m K for glasses used in windows. Atrium building thermal conductivity was considered 1.05 W/m K, glass thickness 30 mm, reflection 5%, absorption 65% and penetration 27%.

The indoor comfort limit temperature value for the heating and cooling load within the building on January 21 and July 21 has been considered to be 25°C in both atrium and courtyard building options. As to the opacity and penetration values outdoors, configurations were made for the glasses used at windows as reflection 10%, absorption 34% and penetration 56%. As data entry; the time zone of the buildings, latitude, longitude, cloudiness ratio as well as month, date, time, minute were included. Thermal comfort and radiation model module have also been used for comfort calculation. The extent of absorption of light by the weather and the data of radiation of light indoors has been included therein. Since these factors are very close to zero, they were considered at 0.001098 levels.

6. Numerical solution modelling

Computational fluid dynamics (CFD) programs are powerful design tools that can predict detailed flow movement, temperature distribution and solar heat flux. Recently, computational fluid dynamics has been used to predict convective heat transfer at exterior building surfaces [1–19, 24–30].

Details of temperature distribution, airflow patterns and other comfort parameters would provide a better picture of the resultant thermal performance within the atrium in response to the changes of building design variables. CFD's main advantages for this application are:

1. Ability to analyse a specific or complex building or building configuration,

2. Ability to acquire very high spatial resolution data,

3. For atmospheric conditions, high Reynolds number flows can be incorporated and

4. Access to detailed information on the flow field and the thermal field.

In these previous studies, this allowed for a detailed analysis of, the correlation of heat transfer coefficient distribution over building surfaces; the influence of turbulence and wind direction; the correlation with different reference wind speeds; the thermal boundary layer, etc. However, some important limitations of the applied numerical models have to be emphasized. Considering the building shell for evaluation of the thermal comfort and energy performance of the courtyard and atrium buildings also studied in CFD. A number of layers with varying physical properties and thicknesses constitute the wall section. The outside surface is subject to radiation exchange ($q_{r,o}$), convection heat transfer ($q_{c,o}$) and solar radiation (I_s) from the sky. The combined convection and heat transfer (q_i) within the inside surface directly defines the extent of air conditioning needed to preserve the intended interior temperature ($T_{f,i}$). The assumptions below were employed in the mathematical model:

- No heat generation is observed.
- Interface resistance is negligible due to good layer contact.
- Negligible thermal properties variation.
- Small composite roof thickness relative to other dimensions.

Therefore it is safe to assume a one-dimensional temperature variation.

Based on daily average wind speed and the direction of heat flow, the convection coefficient is constant. The above assumptions, lead us to a conduction equation which uses the composite roof for directing the heat transfer:

$$\frac{\partial^2 T_j}{\partial x^2} = \frac{1}{\alpha_j}\frac{\partial T_j}{\partial t} \tag{1}$$

Here, 'c' is the specific heat, 'ρ' is the density, k is the thermal conductivity, the subscript j refers to the layer, i.e. j = 1, 2,..., N and α is the thermal diffusivity ($k/\rho c$). For the purpose of obtaining temperature variations and heat-transfer rates subject to prescribed initial and boundary conditions, the problem centres around the fundamental solution of Eq. (1), applied to all layers. Initial temperature was accepted as uniform and equal to outside ambient temperature daily mean values; $T_{f,o,}$ mean.

There are many different correlations in the literature to determine the external heat transfer coefficient for the buildings. Palyvos summarized different correlations found in the literature. On the basis of thirty available linear correlations, Palyvos recommended using the following correlation, Eq. (2) to calculate the heat transfer coefficient (h_c) for windward surfaces: [20]

$$h_c = 4\,V_w + 7.4 \tag{2}$$

where V_w is wind velocity. With zero wind velocity, an external heat transfer coefficient value of 7.4 W/m^2-K was used [20].

The optical properties of the glazed facade surface (semi-transparent), i.e. a solar transmittance of 36% and an absorptivity of 17.5%, were the same as those used in previous studies of an existing atrium building at Concordia University. A single glazed wall with a total overall thickness of 24 mm and an effective thermal conductivity of 0.0626 W/m²-K was used to simplify the modelling of the glazed façade. Also considered were the radiation exchange between the facade and the sky [23].

The correlation was used to calculate the sky temperature [14],

$$T_{sky} = [\, \varepsilon_{sky}\, T_{out^4}]^{1/4} \tag{3}$$

where the emissivity of the sky, ε_{sky}, was calculated using the relation, $\varepsilon_{sky} = 0.727 + 0.0060\, T_{out}$ with an ambient temperature of $T_{out,}$ of 25°C. The heat sources were modelled as a no-slip wall boundary (2.00 × 2.00 m) located in the centre of each floor. In all cases the buoyancy flux value, B, was assumed to be 22.63 × 03 m^4 s^3 (a heat source of 823 W was assumed approximately equivalent to four sitting persons with desktop computers) on each room floor and 14.57 × 103 m^4s^3 (a heat source of 530 W was assumed approximately equivalent to seven resting persons) on the atrium floor. A constant relative pressure of 0 Pa was used across the room inlets and the atrium outlet [30].

Other value can be used in the model since the steady periodic solution is independent of the initial temperature distribution. The boundary conditions are given as follows:

(i) Boundary conditions at the inside surface ($x = 0$):

$$-k_1 \frac{\partial T}{\partial x}\Big|_{x=0} = h_i (T_{f,i} - T_{x=0}) \tag{4}$$

here h_i is the combined heat-transfer coefficient for inside surface; based on ASHRAE handbook of fundamentals [21]: h_i = 9:26, W/m²K for upward direction of heat flow and h_i = 6:13, W/m²K for downward direction of heat flow:

(ii) Boundary conditions at the outside surface (x = L):

$$-k_N \frac{\partial T}{\partial x}\Big|_{x=L} = h_{c,o}(T_{x=L} - T_{f,o}) - \lambda I_s - q_{r,o} \tag{5}$$

here $h_{c,o}$ is the exterior surface convection coefficient, $T_{f,o}$ is the exterior ambient temperature and I_s the outside surface 's solar absorptivity.

The coefficient ($h_{c,o}$) is a function of wind speed (υ). Empirical values are taken from Ito et al. [22] as

$$h_{c,o} = 18.63\, V^{0.605}\ \mathrm{in W/m^2 K} \tag{6}$$

and

$$V = \begin{cases} 0.25\upsilon & \text{if } \upsilon > 2\text{m/s} \\ 0.50\upsilon & \text{if } \upsilon < 2\text{m/s} \end{cases} \tag{7}$$

The temperature ($T_{f,o}$) is determined by a sinusoidal function based on a 24-h period. Here t = 0 corresponds to midnight, as

$$T_{f,o} = T_{f,o,\text{mean}} + A_o \sin(\omega t - \phi) \tag{8}$$

here A_o being the amplitude, ω is the frequency and is the phase. The solar radiation (Is) is calculated for horizontal roofs in atrium and courtyard buildings in Turkey, employing the ASHRAE clear-sky model [23–31]. The nonlinear radiation exchange ($q_{r,o}$) is provided by [30–34, 35].

The nonlinear radiation exchange ($q_{r,o}$) is given by,

$$q_{r,o} = \varepsilon\sigma(T_{sky}^4 - T_{x=L}^4) \tag{9}$$

here σ is the Stefan–Boltzmann constant, ε is the surface emissivity and T_{sky} is the sky temperature and is considered equal to ($T_{f,o}$-12).

The airflow patterns in atrium and courtyard buildings and temperature distributions in the atrium and courtyard buildings are governed by the conservation laws of mass, momentum and energy. The mathematical model applied includes the numerical techniques to solve the continuity, Navier–Stokes (N-S) and energy equations for incompressible, three-dimensional and turbulent flow. The general form of the momentum, turbulent kinetic energy, turbulent energy dissipation and energy (temperature for constant heat capacity) equations in the steady state form can be expressed in the general form as follows:

$$\frac{\partial(\rho u_i \phi)}{\partial x_i} = \frac{\partial}{x_i}\left(\Gamma_\phi \frac{\partial \phi}{\partial x_i}\right) + S_\phi \tag{10}$$

where variable (Ø) is Ø) = (u), (v), (w), (k), (ε), (h, T), respectively, ΓØ is the diffusion coefficient of the variable (Ø) and SØ represents the source terms including the pressure terms, thermal source terms, etc., as appropriate for the variable (Ø) being solved.

6.1. Energy modelling for atrium and courtyard buildings

Heat transfer by thermal radiation, conduction and convection is an extremely important consideration in many modelling cases such as the case being investigated here.

The solar radiation was considered to transmit through the glazing facade wall and heat up interior surfaces of the building, as well as partially absorbed at the glazing facade wall.

In order to determine the temperature distribution in atrium buildings and buoyancy item in energy conservation equation should be solved [32].

$$\frac{\partial \rho T}{\partial t} + \frac{\partial \rho V_i T}{\partial X_j} = \frac{\partial}{\partial X_j}\left[\Gamma_{T,eff} \frac{\partial T}{\partial X_j}\right] + \frac{q}{c_p} \tag{11}$$

where $\Gamma_{T,eff}$ is the temperature effective diffusivity; q the heat source; C_p is the specific heat at constant pressure.

In this research, the equation below is used to estimate the temperature effective diffusivity.

$$\Gamma_{T,\text{eff}} = \frac{\mu_{\text{eff}}}{\text{Pr}_{\text{eff}}} \tag{12}$$

where P_{reff} is the general Prandtl number.

Turbulent effects are united to turbulent effective diffusivity, which is the sum of turbulent diffusivity μ and laminar viscosity coefficient.

$$\mu_{eff} = \mu^1 + \mu \tag{13}$$

In the assumption of Prandtl-Kolmogorov, turbulent diffusivity μ is the result of turbulent fluctuation momentum energy and turbulent fluctuation dimension. l is used to denote turbulent length proportional scale.

$$\mu 1 = Cv\rho\, k^{1/2} \tag{14}$$

where $C_v = 0.5478$ is the empirical constant [33].

7. Modelling and simulation

The typology of examined courtyard and atrium models were depicted and digital mesh networks of each defined atrium and courtyard model were drawn, together with thermal regions of each model, model surfaces were generated including restricting conditions. Geographical and climatic data from various climatic regions were then entered into the Star CCM+ simulation program. In addition, data pertaining to constructional materials, constructional components and structure envelope permeability and reflectivity were entered. The thermal regions, building surfaces and elements thereof were determined beforehand. Later, the data for interior thermal gains was entered and analysis was conducted. The criteria included in optimization studies were; 21st day of July for the cooling period in the summer months and 21st day of January for the heating period, hourly, daily, day and night building temperature and average temperature distributions, inter-building total temperature gain and loss values, air direction, air layering, air change ratio for courtyard and atrium buildings' thermal zones, thermal zones among courtyard and atrium buildings, for all building surfaces and roof area; overall and average heat transition amount, surface temperatures, pressures and velocity distributions, inter-building and especially courtyard 1.60, 3.20 and 6.50 m level horizontal-section temperature, pressure and wind speed values were studied and in regard to such values, average temperature and internal temperature distributions, general temperature gain, total temperature loss and gains of sunlight on the surface of the courtyard and atrium building were estimated. In order to provide better cooling and ventilation throughout the cooling season, inter-building temperature gains and losses throughout the heating season were estimated and evaluations were carried out in order to reveal the effects of such results on cooling and ventilation load. All such values were presented in numerical and visual reports and evaluations were conducted and comments were made on internal temperature and average temperature distributions on the courtyard and atrium building surface, overall temperature gain, total temperature loss calculations as well as their effects on cooling and ventilation, based on these values.

In the CFD software Star CCM+ program where the analysis study is performed, information on the building envelope such the thickness, density, specific heat, thermal conductance coefficient, sun-radiation reflectivity, surface roughness, sun radiation absorbency and number of layers are defined whereas layers in the floorings together with (if present) separate stratifications are defined in the mezzanine floor, ground floor and roof slab. The material used in the flooring was examined in terms of its specific heat, density, thickness, surface roughness, sun radiation absorbency, thermal conductance coefficient, sun radiation reflectivity and number of layers. On the other hand, by entering the values of volume ambient temperatures, boundary conditions for surfaces and thermal zones, absorbency of surfaces, density, reflectivity, thermal conductance and specific heat, we were able to acquire the data used in the simulation program.

8. Result and discussion

8.1. Inner courtyard wind tunnel measurements: wind velocity and distribution results

The aim of this section is to determine air movements and the wind velocity that will occur in the courtyard, along with the turbulence values at the predetermined points and then to compare these values with the performed measurements.

First, before moving to the interior courtyard measurements for the courtyard model, velocity profiles are obtained for windward 3–4 axis over the wind and with '0.00 cm–2.00 cm–4.00 cm –6.00 cm' distances from the building on the building surface over the wind region at the building model placed in the observation room of the wind tunnel (**Figure 8**).

According to the velocity profile diagrams obtained as a result of the measurement, wind velocity reaches to 6.00 m/s at H/2 distance away from the building at approximately H/3 height (**Figure 6**).

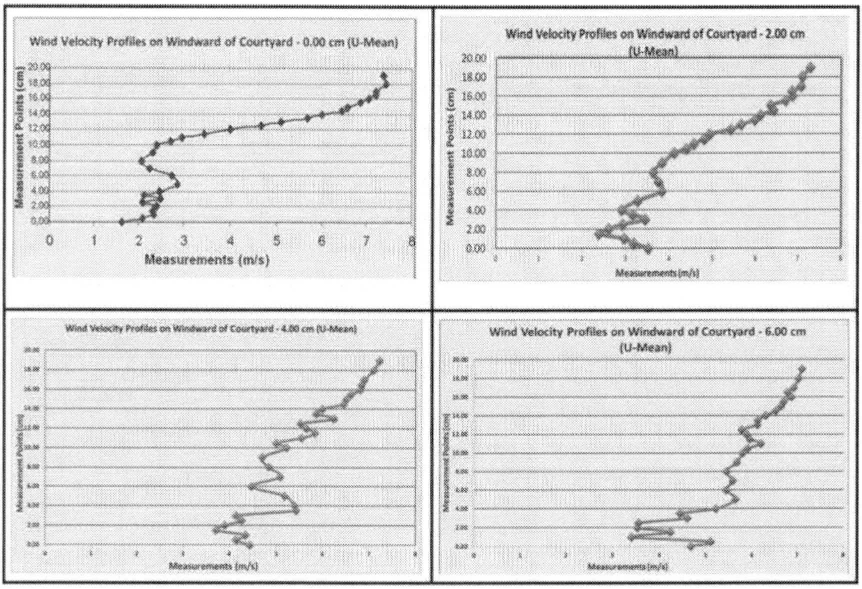

Figure 6. Wind velocity profiles of the windward region of the courtyard typology.

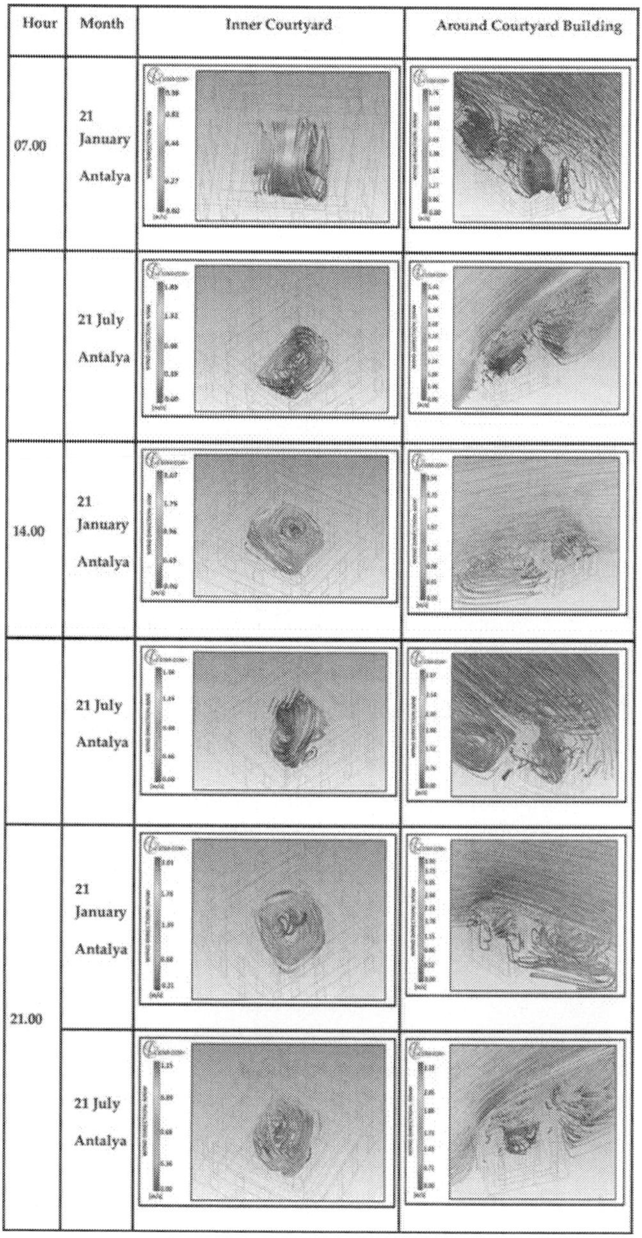

Figure 16. The wind flow and turbulence streamline values for inner courtyard and exterior of courtyard in Antalya for the hot-humid climate region.

City	Antalya (ANT)		Diyarbakir (DB)		Erzurum (ERZ)	
Month	January	July	January	July	January	July
07:00	NNW	NNW	WNW	WNW	WSW	ENE
14:00	SSW	NNE	NNW	ENE	W	ENE
21:00	NNW	NNW	WNW	NNW	SSE	ENE

Table 2. The leading wind vectors for the studied period of 07:00-14:00-21:00 for long-term meteorological mean values for three different climate regions for the courtyard typology.

8.6. The evaluation of inner courtyard thermal comfort as per wind tunnel experiments and CFD numerical analysis

The primary thermal performance of the courtyard depends primarily on the shape feature of the courtyard structure and dependent on the envelope of the structure's envelope; it can show changes related to amount of solar radiation which permeates inside from the opaque surfaces and as per the position of the sun. In order to obtain suitable comfort conditions, different strategies are utilized for different climate regions. However, all of them would create an advantage only when there is comfort required inside and when the solar radiation causes comfort or discomfort. Shading would allow for escape from heat retention in the summer, while avoiding shading in winter in order to retain the heat would be the primary rule.

The geometry of the courtyard form would strongly affect the shaded area ratio created by inner surfaces and as a result, it affects the solar radiation which is received as well as the thermal performance of the building. It is generally seen that deep courtyard forms are recommended in order to gain the maximum interior shaded area in the summer. However, in the winter using the shallow form for obtaining sunny areas would be a better solution. The courtyard ground would have a small effect in the winter on the heat production and heat transfer due to solar radiation; but in the winter this effect is less noticeable.

The study by Taleghani et al. on the effect of different building styles on building comfort and energy performance as well as the study by Yasa et al., on different courtyard types on energy and comfort as depending on the climate zone are the studies which mainly discuss the issue at large [22, 25, 34, 35].

Taleghani et al. made a comparison between different building forms regarding the effect of different building typologies on indoor energy and comfort. They showed, all else being equal, that the larger surface to volume ratio of a courtyard building (and its envelopes), the higher heat loss and consequently energy demand for heating compared to a building with no open space. The current research added a linear building type to these comparisons, using 1-, 2- and 3- storey models [25, 34, 35].

Although it is found that the courtyard dwelling performing with less energy efficiency compared to a building with a square floor plan of the same size, the present study showed that the courtyard dwelling was more energy-efficient. This discrepancy can be understood with reference to the surface to volume ratio of the dwellings with a square floor plan.

From the energy point of view, the energy consumption for heating and lighting of the single and linear shape models increases when the number of floors in the models increases. This amount is slightly decreased for the courtyard shape. This observation also applies to the 2- and 3-storey models (average of all floors). In this case, by increasing the number of floors, it is observed that the average of the number of comfort hours in the single-zone building decreases.

Conversely, the average of the number of comfort hours in the linear and the courtyard shape model increases by increasing the number of levels from one to three.

It is observed that the comfort situation on January 21 for the places which are oriented towards the atrium is at higher values for the surface temperature values as compared to a courtyard dwelling. Hence, in the January 21 heating periods, the atrium typology may be preferred from energy performance point of view. The reason for the optimization for different climates for inner courtyard comfort would be to obtain sufficient sun radiation (heat convection) and to either reduce the need for cooling or to get rid of it completely by creating sufficient shading in the summer. The sunrays (radiation) received by the courtyard is seen as the primary factor which affects the thermal performance of the building. The irradiation ratio usually depends on the location of the building, the climate conditions of the location, the time of the year and the configuration of the courtyard. The solar radiation which is absorbed has the function of increasing the surface temperatures and as a result it will increase the temperature of the nearby air layers. This has an important effect on the thermal conditions of the courtyard empty area. As a result, it is required to allow for maximum amount of solar irradiation to enter the courtyard, so that the thermal performance can be achieved both in summer as well as in the winter. In order to achieve this it is evident that proper configuration and proportioning needs to be done in the inner region of the envelope of the courtyard building.

It is observed by calculating between the sunny areas and the shadowed areas that the self-shading generation of the courtyard building on the building surfaces which are oriented towards the courtyard decreases the sunny area by 4%. However in the winter, this self-shading has an effect of increasing the heating load by 12%. Especially in the cold region Erzurum, it is seen that gaining sun radiation in the winter is more important than avoiding it in the summer.

Building surfaces absorb the sun light as long as they are exposed to it. For many opaque surfaces, the area of surface which absorbs the sunlight depends on the absorbance of the surface. On the surfaces which are exposed to the sun, an agreeable strategy would be to restrict the sunrays before they hit the surface. In the hot summer and in relatively colder climates as well as in very cold and in moderate climates, this concept and the effect of the principles show differences. In the above-mentioned environmental conditions, if special design criteria are applied during the design stage of the building, then it can be an effective environmental regulator.

9. Conclusion

According to the CFD numerical analysis, in hot-dry and hot-humid climate, during the heating season, the differences are clearly visible. The average winter monthly heating demand of building courtyard; it is more than of atrium model (excluding summer month 21 July in which the heating demand is zero). The average winter heating season of the year difference for atrium-building typology is higher than courtyard typology. This situation shows that in hot-dry and hot-humid climate covering the transitional space, thereby creating more transparent surface, this could potentially reduce the heating demand for atrium building. Conversely, overheating risk should be checked for atrium typology, which typically increases the number of summer discomfort hours as shown in **Figure 12.** Atrium typology decreases its annual energy use, but increases the number of discomfort hours in the summer. Contrary, courtyard-building typology increases its annual energy use, but decreases discomfort hours in summer.

The courtyard models have a lower number of discomfort hours and higher heating energy demand in comparison with their atrium models. Therefore, for an atrium typology should be used for heating season (limiting heat losses) of the year, whereas the advantages of the courtyard should be used for summer (reducing overheating). When we look at the simulation results, we can see that it would be efficient if we use courtyard option for 21 July cooling season of the year. Comparing buildings with different climatic regions and seasons of the year, the courtyard typology has the smallest number of summer discomfort hours. This is because of the shading effect of the courtyard and wind effect in courtyard for the surrounding building.

Dealing with solar absorption and ventilation in a courtyard is problematic. The dimensions of a courtyard can influence the quantity of sun and wind allowed or blocked. In summer, when we compare the building typology or form; with regard to comfort and energy consumption; less absorption and more ventilation is favourable. Conversely, more sun and less wind are preferable in winter. In summer, the sun angle is high and a compact form provides more shading while a less compact form allows more sun to penetrate in winter. Likewise, a compact form breaks cold winds in winter but is less ventilated in summer. This result can conclude from E. Yasa's Ph.D. thesis and paper [36]. An efficient design strategy could be based on the weight of the heating or cooling energy consumption. Hence, this shows that the design of a courtyard depends on the policies of energy consumption on a national or regional level.

If we compare the orientation effect on building comfort, we can conclude that the North-South orientation provides the coolest microclimate within a courtyard block for a pedestrian. This orientation keeps a courtyard shaded from the early morning till 2 hours before noon and again 2 hours after noon till sunset. Likewise, the indoor environment of the building absorbs the sun while it has provided shading for the courtyard. This makes the North-South orientation the least comfortable model and East-West the most comfortable model in summer from the perspective of the indoor environment. A square plan (like the 6.00 m × 6.00 m × 6.00 m courtyard) could be a balance that satisfies both a person within the courtyard and a dweller inside the building.

In the warm-dry and in the warm-humid climate regions, it has been observed that due to the wind flow entering the courtyard and due to the shading created from the geometry of the building typology; the formation of comfortable areas takes place, especially during the cooling period. It is observed that in the hot and humid regions, the thermal conditions in the summer nights can reach warm temperature values similar to day values, even though the temperature may be a few degrees lower at night (3 °C to 5°C).

http://dx.doi.org/10.5772/66669

As a result, due to the facts that they contain more complex air patterns and that in order to create the fundamental comfort conditions they have a greater energy consumption and thus the few studies which have been conducted on courtyard- and atrium-type buildings currently are not sufficient to allow generalization of the topic. Hence, the subject matter needs to be investigated further with one on one measurement as well as with the use of both experimental and numerical studies.

Acknowledgments

In this study, numerical simulation section with Star CCM+ was analysed by super computer in Texas A&M University Department of High Performance Super Computing Division.

Author details

Enes Yasa

Address all correspondence to: enesyasa@yahoo.com

Department of Architecture, Faculty of Architecture, European University of Lefke, Lefke, Cyprus

References

[1] Defraeye T., Blocken B., Carmeliet J. 2010. CFD analysis of convective heat transfer at the surfaces of a cube immersed in a turbulent boundary layer. International Journal of Heat and Mass Transfer, 53 (1–3), 297–308.

[2] Bagneid A. 2006. The creation of a courtyard microclimate thermal model for the analysis of Courtyard Houses [online] PhD dissertation. Texas A&M University, Texas, USA.

[3] Medi H. 2010. Field study on passive performance of atrium offices [online]. 1st International Graduate Research Symposium on the Built Environment. held 15–16 October 2010 at Middle East Technical University (METU), Turkey.

[4] Hung WY. 2003. Architectural aspects of atrium. International Journal on Engineering Performance-Based Fire Codes, 5 (4), 131–137.

[5] Goulding JR., Lewis O. and Steemers TC. (eds.) 1993. Energy in Architecture: The European Passive Solar Handbook. London: B.T. Batsford Limited.

[6] Samant S. 2011. A parametric investigation of the influence of atrium facades on the daylight performance of atrium buildings [online]. PhD Thesis.

[7] Taleghani M., Tenpierik M. and Dobbelsteen A. 2012. Environmental impact of courtyards: a review and comparison of residential courtyard buildings in different climates. Journal of Green Building, 7 (2), 113–136.

[8] Saxon R. 1986. Atrium Buildings Development and Design, The Architectural Press, 2nd edition, London.

[9] Mills FA. 1994. Energy efficient commercial atrium buildings. ASHRAE Transactions, 100 (1), 665–675.

[10] Chun et al. 2004. Thermal comfort in transitional spaces- basic concepts. literature review and trial measurement. Building and Environment, 39 (2004), 1187–1192.

[11] Safarzadeh H., Bahadori MN. 2005. Airflow in Buildings with Courtyards. School of Mechanical Engineering, Sharif University of Technology, Tehran, I. R. of Iran. 2005.

[12] Aldawoud, A., 2008. "Thermal Performance Of Courtyard Buildings", Energy and Buildings 40 (2008) 906–910.

[13] Meir I. A., 2000. "Courtyard Microclimate: A hot arid region case study" in Steemers K. & S. Yannas (eds.) (2000) Architecture-City-Environment.Proc. 17th PLEA Int. Conf. Cambridge, UK July 2nd-5th - refereed papers, pp. 218–222.

[14] Dunham DD. 1990. The courtyard house as temperature regulator. New Scientist, 8, 663–666.

[15] Mabb J. 2008. Modification of Atrium Design to Improve Performance: Atrium Building Physics, Berlin: Verlag Dr. Muller.

[16] Aldawoud A., Clark, R. 2007. Comparative analysis of energy performance between courtyard and atrium in buildings. Energy and Building, 40 (3), 209–214.

[17] Ratti C., Raydan D., Steemers K. 2003. Building form and environmental performance: archetypes, analysis and an arid climate. Energy and Buildings, 35, 49–59.

[18] Mohsen MA. 1979. Solar radiation and courtyard house forms II: Application of the model. Building and Environment, 14, 185–201.

[19] Muhaisen, AS., Gadi MB. 2006. Shading performance of polygonal courtyard forms. Building and Environment, 41, 1050–1059.

[20] Rajapaksha I. 2004. "Passive Cooling in the Tropics: A Design proposition for Natural Ventilation", Built environments and environmental buildings: Proceedings of the 21st International conference Passive and Low Energy Architecture, Vol.1, M.H. de Wit (ed), PLEA. 2004, Eindhoven, The Netherlands, September 19- 21, pp. 531–536.

[21] Rajapaksha, I., Nagai H., Okumiya M. 2003. A ventilated courtyard as a passive cooling strategy in the warm humid tropics. Renewable Energy, 28(11), 1755–1778.

[22] Yasa, E., Ok, V. 2014. Evaluation of the effects of courtyard building shapes on solar heat gains and energy efficiency according to different climatic regions. Energy and Buildings, 73, 192–199.

[23] Lomas K.J. 1992. Thermal modelling of building envelopes: the state of the art, in: Proceedings of the Symposium on Energy Efficient Buildings, Kowloon, Hong Kong.

[24] Hussain S., Oosthuizen Patrick H. 2013. Numerical study of buoyancy-driven natural ventilation in a simple three-storey atrium building. International Journal of Sustainable Built Environment, 1, 141–157.

[25] Emmel MG., Abadie MO., Mendes N. 2007. New external convective heat transfer coefficient correlations for isolated low-rise buildings. Energy and Buildings, 39(3), 335–342.

[26] Blocken B., Defraeye T., Derome D., Carmeliet J. 2009. High-resolution CFD simulations for forced convective heat transfer coefficients at the facade of a low-rise building. Building and Environment, 44 (12), 2396–2412.

[27] Palyvos J.A. 2008. A survey of wind convection coefficient correlations for building envelope energy system's modelling. Applied Thermal Engineering, 28, 801–808.

[28] Meir IA. et al. 2004. The vernacular and the environment towards a comprehensive research methodology, The 21th Conference on Passive and Low Energy Architecture. Eindhoven, the Netherlands, 19–22 September 2004.

[29] Ito N., Kimura K., Oka, J. 1972. IIA field experiment study on the convective heat transfer coefficient on exterior surfaces of a building. ASHRAE Transactions, 68: 184.

[30] Mustafa AF., Costa F.J. 1983. Al Jarudiya: A model for low rise/high density developments in the Eastern provinces of Saudi Arabia, in Islamic Architecture and Urbanism, ed. A. Germen, University of Dammam, Saudi Arabia, pp. 239–256.

[31] Brarozzi GS., Imbabi MSE, Nobile E., Sousa ACM. 1999. Physical and numerical modelling of a solar Chimney-based ventilation system for buildings. Building and Environment, 27(4), 433–445.

[32] Qingyan, Chen, Leon, R., Glicksman. 2001. Simplified methodology to factor room air movement and the impact on thermal comfort into design of radiative, convective and hybrid heating and cooling systems. ASHRAE RP-927, ASHRAE.

[33] Taleghani M., Tenpierik M., van den Dobbelsteen A., de Dear. R. 2013. Energy use impact of and thermal comfort in different urban block types in the Netherlands. Energy and Buildings, 67 (0), 166–175.

[34] Taleghani M., Tenpierik M. and Dobbelsteen A. 2013. The effect of different transitional spaces on thermal comfort and energy consumption of residential buildings. in 7th Windsor Conference: The changing context of comfort in an unpredictable world Cumberland Lodge. 2012, Network for Comfort and Energy Use in Buildings, Windsor, UK.

[35] Yasa E., Fidan.G, Tosun M., 2014. Analysis of historic buildings in terms of their microclimatic and thermal comfort performances "Example of Konya Slender Minaret Madrasah". Journal of Architectural Engineering Technology, 3 (3), 2168–9717. doi:10.4172/2168-9717.1000126.

[36] Yasa, E. 2010. A method developed in the optimization of courtyard building shape according to climatic performance in different climate regions. Ph.D. thesis. Institute of Science and Technology, Istanbul Technical University, February 2010.

Chapter 5

Energy-Efficient Building Design in the Context of Building Life Cycle

Izzet Yüksek and Tülay Tikansak Karadayi

Additional information is available at the end of the chapter

http://dx.doi.org/10.5772/66670

Abstract

Energy is one of the most important catalysts in wealth generation, economic growth, and social development in all countries. Buildings have a significant share in total energy consumed globally; therefore, they have a profound impact upon the environment. Energy is used in every stage of building life cycle (these stages are choice of locality, architectural design, structural systems and material selection, building construction, usage and maintenance, demolition, reuse-regain-recycle, and waste disposal). According to World Watch Institute data, buildings are responsible for the annual consumption of 40% of the world's energy. Energy consumption of buildings can be reduced significantly in every stage of a building life cycle. This study investigated the energy-efficient methods in building life cycle. In this context, we give information about the life cycle of building and explain energy-efficient guiding principles in life cycle stages.

Keywords: building—life cycle, building—energy consumption, energy efficiency in buildings, energy and environment

1. Introduction

Buildings consume energy at different levels in every stage of life cycle. Approximately half of all nonrenewable resources (water, energy, and raw materials) mankind consumes are used in construction. Contemporary human civilization depends on buildings and what they contain for its continued existence, and yet our planet cannot support the current level of resource consumption associated with them [1].

Construction also has a major impact on the environment in its consumption of energy. For example, building materials occupy a great share of this consumption. The large bulk of materials used consume a great deal of energy for transport [2].

There is a growing concern about energy consumption in buildings and its possible adverse impacts on the environment. These are issues that the building professions in the whole world have to address [3].

Energy consumption is rapidly increasing due to the increase in population and urbanization. Residential energy requirements vary from region to region, depending on climate, dwelling type, and level of development [4]. The construction activities consume 38% of the globally used energy every year [5]. There is a growing concern about energy consumption in buildings and its possible adverse impacts on the environment. These are issues that the building professions in the whole world have to address [6].

Buildings consume energy at different levels and different aims in every stage of the life cycle. In an operating phase, a building with at least a 50-year lifespan, energy used for production of materials, transportation, and construction, "at least five times" as is required in the amount of energy use and operating phases. A large part of the energy (35–60%) is used for heating, air-conditioning, ventilation, and artificial lighting at this stage. Energy-efficient approaches that have the potential to significant energy economy, most of the buildings if you live a long time considered more than 50 years. Even if only to focus on the use and operation phase is very important [7].

2. Methods for ensuring energy efficiency in buildings

It is not possible to bring recommendations of solution that can procure energy efficiency for all buildings. As the function, system, position, and importance of a building changes from building to building, the ways of solution providing energy efficiency will also change. Therefore, a conscious approach needs to be developed in order to reach the right solution at the stage of architectural design through enabling necessary data. In the end, the product to be obtained must be aimed to have the quality of being more efficient, in other words, spending less resource within a longer period of time to perform the same action.

There are very different applications targeting the decrease of energy consumption of buildings. Considering energy consumption in each phase of structuring is achieved with the analysis of building life cycle.

In this respect, we need to know the life cycle of building. Building life cycle is divided into three main phases such as the prebuilding phase, building phase, and postbuilding phase. These phases have some processes. Prebuilding phase includes the appropriate site selection, site planning, building form, building plan, and appropriate space organization, building envelope design choosing energy-efficient building materials, energy-efficient landscape design, obtaining raw materials for building material, manufacturing, and transporting them. The building phase includes the construction and usage processes of the building. The postbuilding phase is the phase following the completion of building usage. In this phase, we have the demolition, recycling, and wipe-out of the building.

The methods applied so as to fulfill the energy efficiency of buildings depending upon life cycle phases.

2.1. Energy-efficient designing methods in the prebuilding phase

The prebuilding phase includes the choice of the space where construction is to be built, the design of the building, the choice of building materials, obtaining raw materials for building material, manufacturing, and transporting them. In these processes, the strategies have been explained with significant energy saving in building life cycle such as the appropriate site selection, site planning, building form, building plan and appropriate space organization, the design of building envelope, the choice of building material, landscape design, and benefitting from renewable energy resources in sequence. These strategies are explained below.

2.1.1. Appropriate site selection

The locations of the hemisphere, slope, and aspect are important design parameters. Location of the building determines the microclimate conditions which has very important role in building energy efficiency, as it is important for learning, climatic values such as sun radiation, air temperature, air circulation, and humidity, which effect energy costs [8].

The site of building and distance between other buildings are one of the most important design parameters, which affect sun radiation amount and air circulation velocity around the buildings. For this reason, the site of the building in the area should be determined to benefit and defend from the renewable energy resources like sun and wind [8].

In order to provide adequate protection from the prevailing wind and sun, the orientation of buildings on the land needs to be appropriate to the climatic conditions of the region. In cold regions, lower overnight temperatures cause colder, denser air to accumulate in hollows and valleys. Therefore, in cold regions it is advisable to position buildings on hillsides rather than in valleys. Such sloping areas are not affected by cold wind as much as valleys and benefit from more direct as shown in **Figure 1** [9].

The topography of the location of the building is important because of the effect the angle of incidence of solar radiation, slope, and orientation of the land in terms of the use of daylight and natural ventilation, solar radiation. If the settlement will be sloping, gained solar radiation energy is reduced in summer, and gained solar radiation energy increases in winter. Therefore, the slope of the land, the amount of incoming solar radiation, and the latitude are very important parameters [11].

It is well known that a south slope is warmer and has the longest growing season in the northern hemisphere. When a choice of site is available, a south slope is still the best for most building types. In the winter, the south slope is the warmest land due to two reasons: the south slope receives the most solar energy on each square foot of land because it most directly faces the winter sun. The south slope will also experience the least shading because objects cast their shortest shadows on south slopes [12].

Figure 2a illustrates the variation in microclimate with different slope orientations. The south slope gets the most sun and is the warmest in the winter while the west slope is the hottest in the summer. The north slope is the shadiest and coldest, while the hilltop is the windiest location. Low areas tend to be cooler than slopes because cold air drains into them and collects there [12].

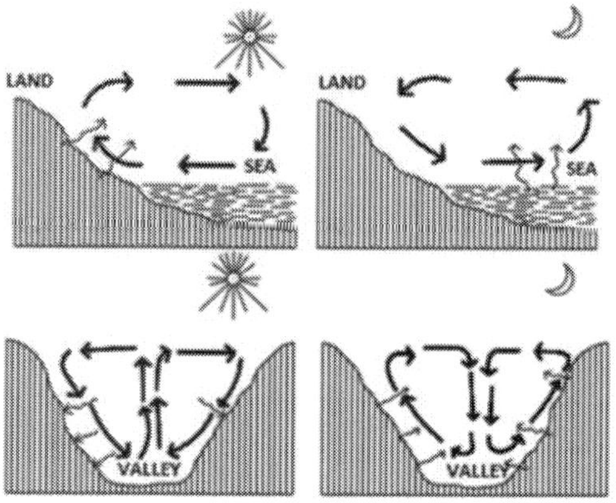

Figure 1. The change of climate conditions surrounding the building depending on the location of the building [10].

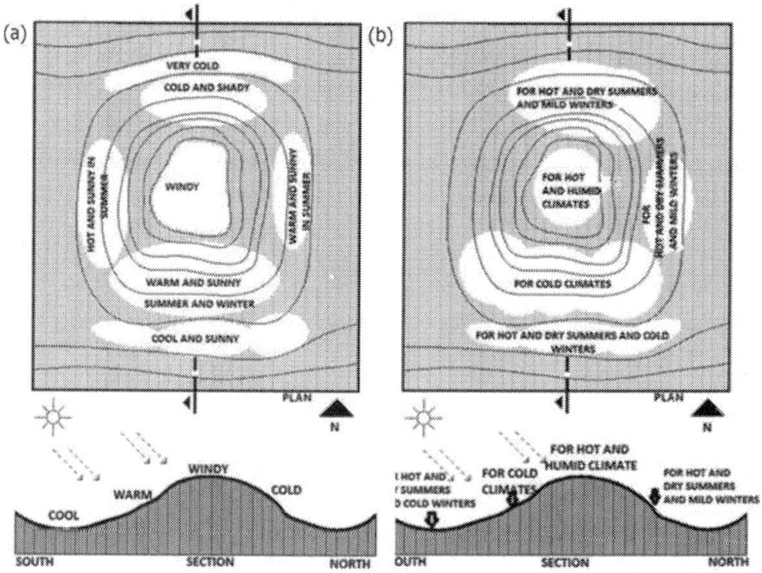

Figure 2. Suitable land for settlement according to the different climate zones [12]. (a) Microclimate around a hill; and (b) preferred building sites around a hill in response to climate for envelope-dominated buildings.

The best side site for a building on hilly land depends on both climate and building type. For envelope-dominated buildings, such as residences and small office buildings, the climate would suggest the sites as shown in **Figure 2b**.

For example, in cold climates, south slopes maximize solar collection and are shielded from cold northern winds. Avoid the windy hilltops and low-lying areas that collect pools of cold air. In hot and dry climates, build in low-lying areas that collect cool air. If winters are very cold, build on bottom of south slope. If winters are mild, build on the north or east slope, but in all cases avoid the west slopes. In hot and humid climates, maximize natural ventilation by building on hilltops but avoid the west side of hilltops because of the hot afternoon sun. Also the cool low-lying areas are appropriate especially to the north of hills. For internally dominated buildings, such as large office buildings that require little if any solar heating, the north and northeast slopes are best [12].

In **Figure 2**, according to the information described above, according to the different climate zones, appropriate residential areas are shown on a theoretical terrain. In terms of climatic effects, the part of the slope that has the mildest qualities for each slope is defined as "thermal belt" [13].

Building altitude leads to differentiation of solar radiation values. As we go above the sea level, we get an increase in solar radiation values. The reason for this increase is dealt with atmospheric conditions, clarity of the atmosphere and shortening of the distance taken. In return, for the increase in solar radiation values, as we go above the sea level, we get a decline in the air temperature. With the increase of altitude, gale force also increases, which leads to the increase of heat loss in the building [14].

2.1.2. Site planning

In the design of buildings, distance between buildings is an important designing parameter that affects utilization of solar energy, wind direction, and speed concerning artificial environment. In the design process, building should be handled as a whole with its environment. The distances between buildings highly affect the energy performance in the usage phase of building. The fact that a building remains within the shading space of other buildings influences the utilization of solar rays and will raise the consumption of energy. In order to utilize solar radiation, building spaces must not be less than the tallest shade height of other buildings. Besides, the position and distance of other buildings affect the speed and direction of wind on building, and this impacts the energy performance of building [15].

Orientation of building affects the ratio of the solar radiation gain of building sides, consequently the total solar radiation gain of building. In addition, the side of buildings affects wind amount, consequently, affecting natural ventilation possibility and heat loss amount by convection and air lack. For this reason, according to the necessities of that region, buildings must be oriented for avoid of or benefit from the sun and wind according to the conditions [8].

As the positioning of buildings as attached to each other would decrease the building envelope/volume rate, declines heat loss and gains through building envelope. In addition, positioning them in the direction of south, southeast, and southwest as an external curve crescent make them utilize solar ray more [16].

In London-BedZED settlement, separate houses having their own gardens were designed. So as to lessen the heat loss of buildings, both compact forms were used and construction groups were gathered together and it tried to decrease the outer surface space/volume rate as shown in **Figure 3** [17].

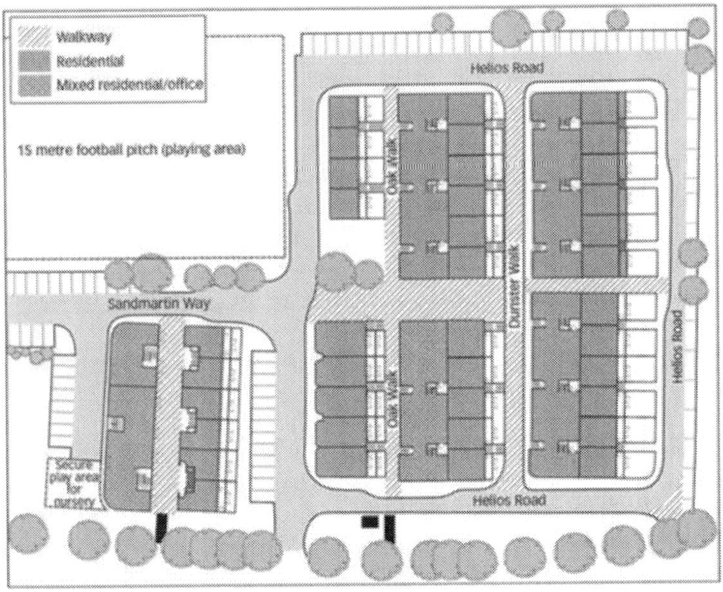

Figure 3. Site plan of London-BedZED ecological settlement [17].

2.1.3. Building form

The shape of building which is a considerable factor affecting heat loss and gain can be defined through geometrical variables making up building such as the proportion of building length to building depth of the building in the plan, building height, type of roof, its gradient, front gradient, and bossages. Heat loss-gain of building may rise and decline depending upon the proportion of the surfaces constituting environment to volume [18].

Energy performance of building is affected by such factors as its form, volume surface rate and frontal motions. There is a direct relationship between the geometrical shape and energy performance of building.

In the conducted studies, it was observed that different results were obtained in the energy performance of the masses which had the same volume but made in different forms [14]. It was calculated that the surface area of the masses has the same volume but different forms. The surface of the cube that was taken as 100 was accepted as a reference (**Figure 4**).

The shape of building is important in areas that have different climate conditions. In cold climate regions, compact forms should be used which minimize the heat loss part. In hot-dry climate regions, compact forms and courtyards should be used which minimize heat gain and helps to provide shaded and cool living spaces. In hot-humid climate region, long and thin forms whose long side oriented to the direction of prevailing wind makes possible maximum cross-ventilation. In mild climates, compact forms, which are flexible more than the forms used in cold climate regions, should be used [8].

http://dx.doi.org/10.5772/66670

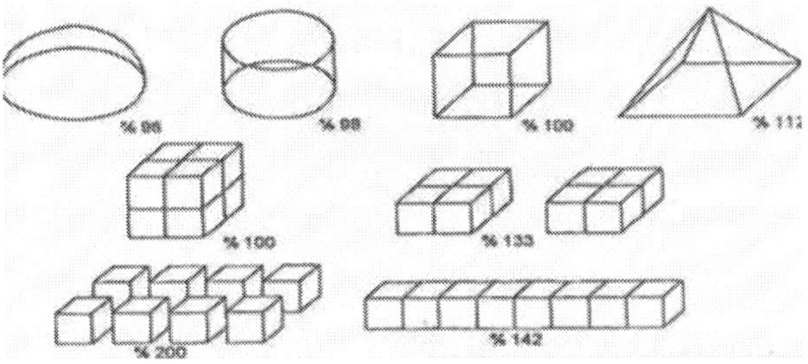

Figure 4. Building form-surface relationship [19].

2.1.4. *Building plan and appropriate space organization*

Building plan and shapes should be effective in energy conservation. Therefore, buildings should be formed to ensure minimum heat gain in warm seasons and maximum in cold. Due to simple plan types such as square or rectangle having a reduced surface area, their heat-loss and -gain are also reduced. Smaller buildings, where internal space has been used efficiently, use less energy as they can be heated, cooled, and illuminated more efficiently than larger buildings [20].

According to the results of the research called "Construction and Energy" performed by German Ministry of Research and Technology, the place of space in the organization of plan is more efficient than the orientation of space with respect to energy consumption [21]. The energy requirement of buildings can be reduced by the internal layout of the design. By making the best use of the sun's radiation, the need for heating energy can be reduced. These communal areas require more heating, whereas spaces with a lower heating requirement such as the pantry, bathroom, and toilet can be used as buffer areas, reducing heat transfer to the exterior by placing these in areas of heat-loss. Spaces such as sun rooms, if located on south façades of buildings, also contribute to heating of the building and energy conservation, by storing solar radiation [20].

In the building design, stratification can perform zoning depending on buffer zone, sanitary spaces, noise level, lighting level, and heating need. Therefore, areas with many users and which are used throughout the day should face southerly direction. Thermal zoning and the settlement of indoors can be designed in a way to raise mutual air motion (**Figure 5**). Deep plans and the use of too many dividing elements may restrict air motion in environments [22].

2.1.5. *Building envelope*

Building envelope is the components such as wall, ceiling, ground, window, and door which separate building (conditioned space) from outdoor and let heat energy transfer into inside or outside. As an indoor and outdoor reagent, it has a vital impact on energy consumption [10]. While the cost of constructing a building envelope makes up 15–40 of the total constructional cost, its contribution to life cycle costs especially to energy cost is around 60% [12]. The skin of building performs the role of a filter between indoor and outdoor conditions, to control the intake of air, heat, cold, and light [24]. Building envelope should minimize the heat loss in the winter and the heat gain in the summer.

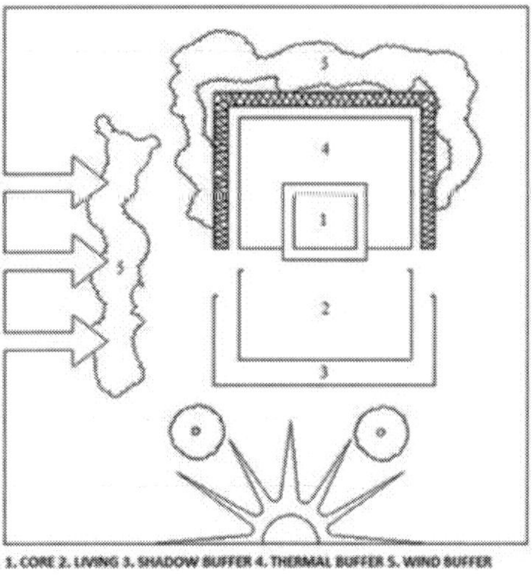

Figure 5. Spatial zoning [23].

The physical and structural specifications of building components, such as walls, windows, flooring, and doors, which make up the outer shell of the building, have a significant impact on the energy consumption of the building. The thermal performance, thickness, and color of the materials used in these components play a significant role in regulating the heat loss and gain of the building [20]. The energy-saving features of the building components analyzed are described below.

Outer walls: Thermal and massive characteristics of outer walls are related to building material constituting them and the characteristics of building element layers and how they are sorted. The walls that will minimize heat loss and gain are well-isolated massive walls with high heat-storing capacity.

The formation of outer surfaces that can get most solar radiation or be protected from radiation in terms of heat gain should be handled depending upon the characteristic of climatic zones. To keep sunlight as much as possible in winter, wall-to-window ratio is desired not to exceed 15% with the use of dark and high-density materials in the parts exposed to the sun [25].

Roofs: In commercial and institutional buildings, roofs are generally flat, and the insulation can be resting on the suspending ceiling. In gabled roof construction where the attic is not used, the insulation is generally in the ceiling [12]. The shape, material, gradient, orientation, outer surface color, and insulating qualities of the roof determine the thermal performance of the buildings. Therefore, roofs need to be designed in such a way to suit the climatic conditions [20].

Thermal insulation qualities of roofs, their gradient and facade should be chosen properly to climatic character, their outer surface color and stratification order should, however, be chosen taking heat gain and loss into account. In temperate dry and temperate humid climatic zone and cold climatic zones, the well-isolated gradient roofs should be preferred. In hot and dry climate zones, flat roofs should be preferred to reduce the impact of solar radiation; in hot and humid climates that allow air flow, raised or sloping roof should be arranged [9].

Windows: Windows affect energy efficiency in buildings via heat loss or gain, natural ventilation, and illumination. The most appropriate direction is south in terms of heat gain, after the east and west side. Large windows reduce the need for artificial lighting while improving daylight [26].

Windows should be designed in the magnitude that is sufficient to provide natural lighting. For example, window magnitude should be at least 15% of the room's floor area [27].

While taking a decision on the transparency rates in building envelope, in which climatic zone the building is placed should be ascertained in advance. Since protection from solar radiation and wind is the basic purpose in hot and arid climatic zones, small and few windows should be used. In hot and humid climatic zones, by taking necessary precautions, large openings should be used in order to raise indoor air circulations. In cold climatic zones, to minimize the heat losses stemming from windows, again small and few windows should be used. Yet, so as to utilize the beneficial effect of solar radiations, the window openings in the southern front should be kept more than the ones in other fronts. In temperate climatic zones, however, it should be given to openings that would enable sufficient air circulation [28].

The use of windows also serves a number of essential purposes such as ventilation, natural lighting, and opening to scenery; it does not bring much load on constructional cost. In the climatic zones having cold winters, positioning window openings in the north should not be preferred due to the fact that heat gain from the sun is too little to be considered and air penetrations increase because winter winds usually blow from the north and thus heat losses grow. It is possible to obtain a certain amount of sun gain from the openings placed in the east and west, even if it is less in winter than the southern front. However, since the summer sun comes horizontally in the morning and afternoon hours, it is very difficult to protect these openings and we may face the problem of overheating. The windows looking toward south, however, may utilize solar rays coming horizontally in winter almost the whole day; in summer, they may be easily protected from the rays coming more vertically [29].

Because of all of these components, southern windows are the systems which can be very commonly used in utilizing sun passively. Yet, compared with wall, due to their weak isolation qualities they are much more open to heat-loss and gain; therefore, it is needed to take precautions for winter and summer. In this case, the application of double-glazing gains a high importance. Night isolation applications, however, are necessary to dismiss the heat losses that may occur after sunset. These isolation elements may be shutter, roller blind, or jalousie fixed either from inside or outside. Or, losses should be reduced through at least bringing curtains strictly down. In summer days, windows may be easily protected by the help of eaves, sunshade, or curtain [29].

In the front, high performance glass that has the most suitable thermal and light transmittance coefficient for the desired qualities depending on climate, sun direction, and the usage purpose of building should be used. Energy can be efficiently used thanks to isolated joineries, low-E covered glasses, argon or krypton-filled double-glazing and air proof detailing and montage.

Doors: The position of outer doors should be chosen considering wind effects, heat gain, and losses. In cold climatic zones, windbreak is suggested in order to be protected from the wind effect increasing heat losses. In hot-arid and temperate climatic zones, as wind does not have a restorative impact on comfort, surfaces closed to wind should be preferred [9].

Floors: Floorings grounded on soil should be arranged in a way to enable the desired performance in terms of heat and moisture. In cold and temperate climatic zones, well-isolated floorings should be preferred. In warm-humid climatic zones, however, heightened floorings can be preferred since air streams become important [9]. In the volumes getting sunlight, floor laying can be used as a thermal heat store. In floor laying, dark color materials having a high heat-storing capacity should be preferred. Not laying carpets on floor and leaving it open increase its capacity of heat absorption.

2.1.6. *Choosing energy-efficient building materials*

Building materials both in the production phase should have energy-efficient features in the use phase. Energy-efficient building material properties are described below.

Local material: In the total energy consumption of constructions, the amount of energy spent for transportation of the construction materials to construction sites is considerable and also affects the constructions' energy efficiency and economical cost. For this reason, if the construction materials are local material and are manufactured in nearby places to the construction site as much as possible, energy consumption in transportation will decrease and that saving in transportation will give the construction an important ecological quality [30].

Recycled resources: A large amount of energy is used in manufacturing many building materials. In the manufacture of building material, using recycled sources instead of the sources which are not newly processed material provides a considerable preservation of raw material and also a considerable amount of energy saving. Recycling building materials are essential to reduce the embodied energy in the building; for instance, the use of recycled metal makes considerable energy savings between the rates of 10 and 90% comparing the material produced from natural resources [31].

Materials manufactured through low density industrial processes: Building materials play a significant role in the energy efficiency of buildings. A large proportion of the total energy used during the building life cycle is consumed during the production of building materials (especially embodied energy). The proportion of the energy amount consumed in the manufacture of construction materials to the total energy amount of a construction with a 50-year process of use consumes in its life cycle processes varies between 6 and 20% depending on the construction methods, climate, and similar conditions [31]. The intensity of energy consumption in the first of these phases for the production of buildings and their components has increased with industrialization [32]. Nonexistence of heavy procedures in the manufacturing process will cause less energy consumption, which provides energy efficiency to materials. Using the developed technologies in industrial processes such as a heat recovery method reduces energy consumption. For instance, in cement manufacturing technology, using the shaft furnaces instead of the conventional rotary furnaces makes energy saving between 10 and 40%. Similarly, the use of an arc furnace instead of a rotary furnace in the steel industry makes about 50% energy saving [31].

Natural materials are quickly obtained from renewable resources: Generally, the energy content of natural materials is lower than that of artificial materials since these materials are manufactured with less energy and labor cost. Such kinds of materials which are easy to be locally provided are generally among the renewable resources. Such vegetal materials used in constructions for instance, wood, bamboo, reed, straw, rye stalk, sunflower stalk, mushroom are the natural materials which are quickly gained from renewable sources [33].

Labor intensive materials: Using highly qualified man power in manufacturing materials will reduce the processes based upon industry, and accordingly decrease the energy consumption. Materials manufactured by using renewable energy resources: especially renewable energy resources (solar energy, wind energy, etc.) instead of fossil fuels should be preferred as a primary energy supplier in the manufacturing process. For example, the adobe brick is dried using solar energy after it is molded [33].

Materials consuming less energy during the worksite process: The management of worksite, the need for electricity energy, and machines in operation, heating, and lightening affect the energy consumption of the worksite. As a result of the increase in mechanization in worksites, the electricity consumption has increased considerably as well [31].

Use of durable building materials: Use of durable materials in the buildings makes them more resistant and long-lasting against various factors. This delays or eliminates the need of renewing material or maintenance due to impairment and aging. In this way, it is saved from the energy spent for the material to be used in maintenance or renewing [33].

Building materials with high thermal insulation capacity: With the choice of building materials whose thermal insulation capacity is high, the energy amount that the construction consumes in its usage stage will be decreased. As mentioned as examples are opaque and translucent insulating materials [33].

2.1.7. *Energy-efficient landscape design*

Through an accurate and conscious energy protected landscape design, it is possible to reduce the energy cost spent for heating and cooling during summer and winter seasons at 30% [34].

The ground flooring of outdoor and grass has a cooling impact via vapor transportation. The materials harboring heat in its body such as asphalt continue to expand heat following sun and they increase night time radiations. So as to reduce the cooling costs spent, using such materials that store heat and reflect lights little or shading them against direct solar rays are among the precautions to be taken [34].

The energy conserving landscape strategies depend on a region. These landscaping strategies are listed by the region and in order of importance as shown in **Figure 6**.

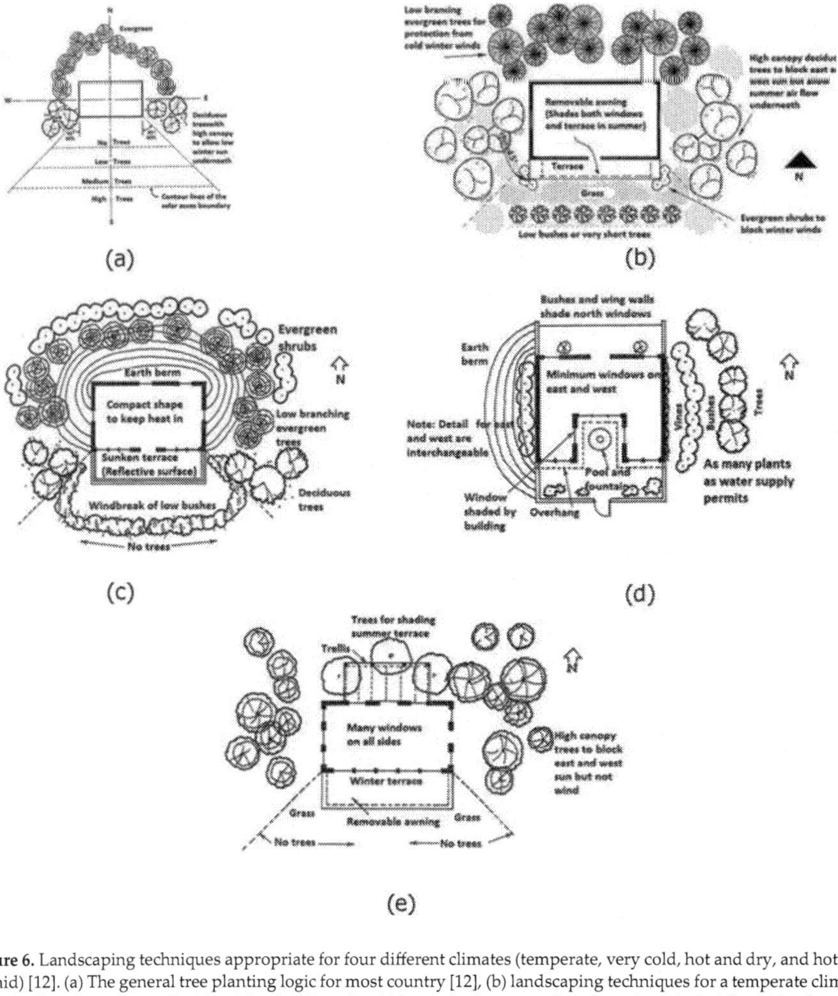

Figure 6. Landscaping techniques appropriate for four different climates (temperate, very cold, hot and dry, and hot and humid) [12]. (a) The general tree planting logic for most country [12], (b) landscaping techniques for a temperate climate. The windbreak on the north side of the building should be no farther away than four times its height, (c) landscaping techniques for very cold climates, (d) landscaping techniques for hot and dry climates, and (e) landscaping techniques for hot and humid climates.

Temperate climate: It should maximize warming effects of the sun in winter and maximize shade during the summer. Buildings should be protected away from winter winds. Summer breezes should be directed toward the buildings. Constantly green trees with low branches to protect them from the cold winter winds on the northern front, low shrubs or trees not high, should be applied on the south front, high body deciduous trees should be placed on the eastern and western facades for block the sun and allowing natural ventilation [12, 35].

Hot-arid climate: It provides shade to cool roofs, walls, and windows. Allows summer winds to access naturally cooled homes and blocks or deflect winds away from air-conditioned homes. North and south sides should avoid forestation, while the eastern and western direction (positioning studies may be substituted), shrubs, vines have been placed on the walls and deciduous trees should be implemented [12, 35].

Hot-humid climate: Channel summer breezes toward the home. Maximize summer shade with trees that still allow penetration of low-angle winter sun. Avoid locating planting beds close to the home if they require frequent watering. Should avoid forestation on the southern front, in the northern front, forestation should be done providing the shadow effect in summer. The eastern and western direction, shrubs, and vines have been placed on the walls and deciduous trees should be implemented [12, 35].

Cool climate: Use dense windbreaks to protect the building from cold winter winds. Allow the winter sun to reach south-facing windows. If summer overheating is a problem, shade south and west windows and walls from the direct summer sun. The north façade is useful in cold climate regions partly raised land application. Northern, eastern, and western fronts in constantly green shrubs and the green, the low branches of trees should be preferred. In the southern wind breaker, low shrubs and grass should be applied. In southeast and southwest direction away from the building, deciduous trees should be used [12, 35].

The ground cover may also be utilized for energy conservation in buildings. Completely or partially buried, construction can moderate building temperature, save energy, and preserve open space and views above the building [36]. If the wall and roof being covered by a layer of earth of substantial thickness sufficient to insulate the dwelling thermally and acoustically and reducing the quantity of energy necessary to maintain the interior of the building comfortable for the occupants even when the atmosphere is extremely hot or cold.

2.1.8. Usage renewable energy resources

Renewable energy sources (sun, wind, biomass, biogas, geothermal energy, hydro, wood, ocean thermal, ebb and flow, wave, sea flows) are the energy resources that can be used by all living creatures on the earth and accepted as inexhaustible thanks to their continuous renewal. It is possible to benefit from renewable energy resources with passive and active methods.

Usage renewable energy resources with passive techniques:

Passive heating: Passive solar heating systems are categorized by the relationship between the solar system and the building. There are three categories of passive solar heating systems: direct gain systems, indirect gain systems, and isolated gain systems [37]. In the passive solar heating system, building elements (windows, walls, floors etc.) collect and store heat and then distributes indoor space.

Direct gain systems: The direct gain passive solar building has windows that admit the winter sun directly into the occupied space. These solar gains serve to either meet part of the current heating needs of building or are stored in the thermal mass to meet heating needs that arise later. Most direct gain buildings include: (1) large, south-facing windows (for north hemisphere) to admit winter sub; (2) thermal mass inside the insulation envelope to reduce temperature swings; (3) calculated overhang above the south glass (or other strategy) to shade the glass in the summer while admitting lower angle winter insolation; (4) a means of reducing heat loss at night. In a direct gain building, sunlight is admitted directly to interior through glazing. It strikes massive interior surfaces (typically concrete floor and masonry wall surfaces), is absorbed, and is converted into the heat. Some of the heat from the surfaces is immediately released back into the room interior. The remainder of heat absorbed is conducted into the thermal mass which slowly warms up; later at night, the stored heat is released back to interior as shown in **Figures 7** and **8** [39].

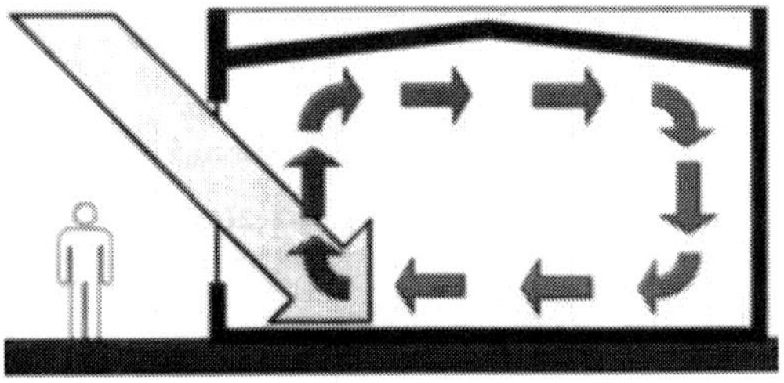

Figure 7. Direct gain schematic [36].

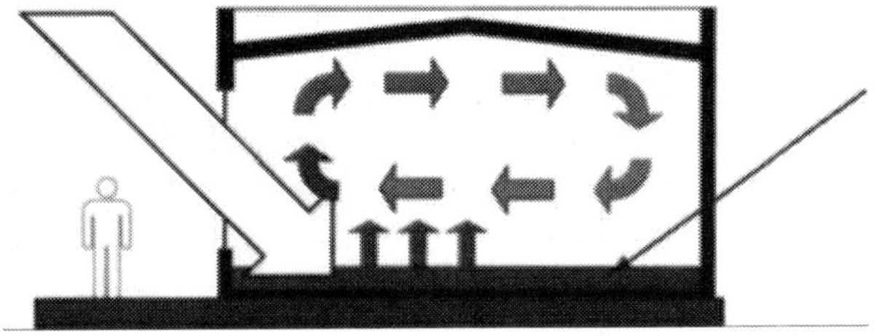

Figure 8. Direct gain plus storage schematic [37].

Indirect gain systems: An indirect gain passive solar system has its thermal storage between facade and the indoor spaces. Heat is collected and stored in an exterior wall or on the roof (with water or brick/concrete) of a building, and distributed to the indoor as shown in **Figure 9** [37].

Isolated gain systems: Isolated gain passive solar concept contains solar collection and storage that are thermally isolated from the indoor space of the building. The most common use in isolated gain systems is a sunspace. Collection and storage are separate from the occupied spaces but directly linked thermally. A sunspace is a room attached to or integrated with the exterior of a building in which the room temperature is allowed to rise and fall outside the thermal comfort zone, as shown in **Figure 10** [37].

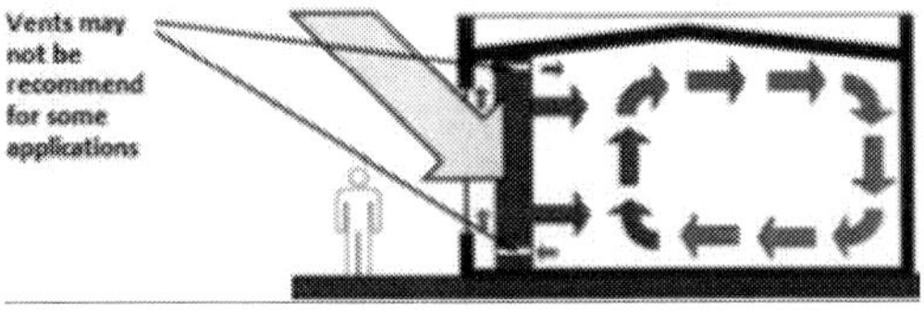

Figure 9. Indirect gain schematic [37].

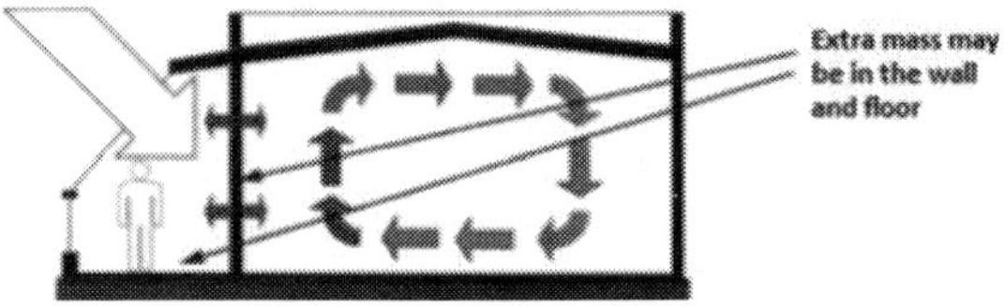

Figure 10. Sunspace schematic [37].

Passive cooling and ventilation: Passive solar heating is divided into categories according to application configuration. On the other hand, passive cooling is better understood as a series of research fields that focus on the basic heat sinks. While this organization is helpful to scientists and inventors, it is a source of frustration for designers and policy makers because so many workable systems involve multiple heat sinks [38]. Nonetheless, this characterization of passive cooling will be described below.

Ventilative cooling: Warm building air and replacing it with cooler outside air. Directing moving air across occupants' skin to cool by combination of convection and evaporation. In passive applications, the required air movement is provided either by wind or by stack effect. In hybrid applications, movement may be assisted by fans, as shown in **Figures 11–13** [39].

Radiant cooling: All building objects radiate and absorb radiant energy. Building objects will cool by radiation if the net flow the outward. At the night, long wave infrared radiation from a clear sky is much less than the long wave infrared radiation radiated from a building. Thus, there is a net flow to the sky, as shown in **Figure 14** [12].

Evaporative cooling: Water has been used to improve the thermal comfort of buildings with or cascades. Because when water is evaporates, energy is lost from the air and reducing the temperature. When water evaporates, it draws a large amount of sensible heat from its surroundings and converts this type of heat in the form of water vapor. As sensible heat is converted to latent heat, the temperature decreases. This phenomenon is used to cool buildings in two different ways. If the water evaporates in the building or in the fresh air intake, the air will be not cooled, but also humidified. This method is called direct evaporative cooling. If, however, the building or indoor air is cooled by evaporation without humidifying the indoor air, the method is called indirect evaporative cooling, as shown in **Figure 15** [12].

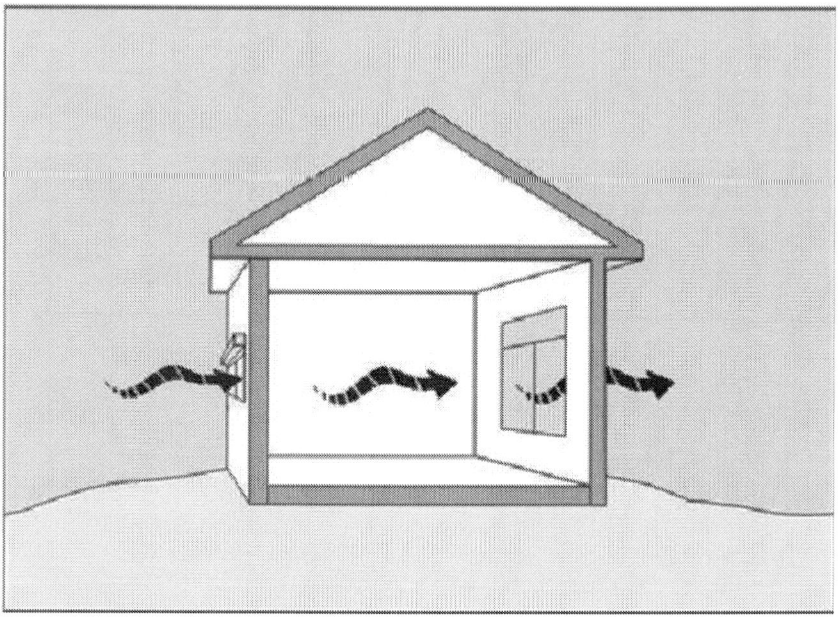

Figure 11. Use windows and doors for cross-ventilation [40].

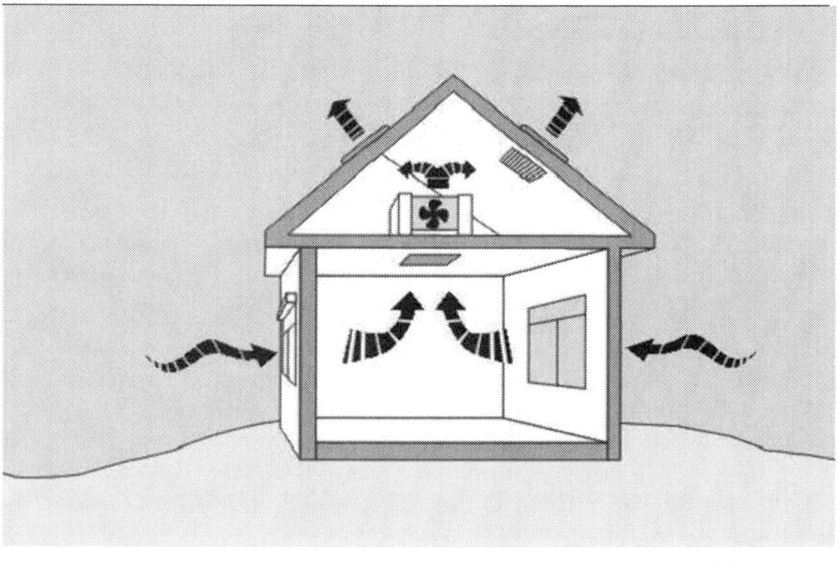

Figure 12. A whole-house fan [40].

http://dx.doi.org/10.5772/66670

Figure 13. Air movement in stack ventilation [41, 42].

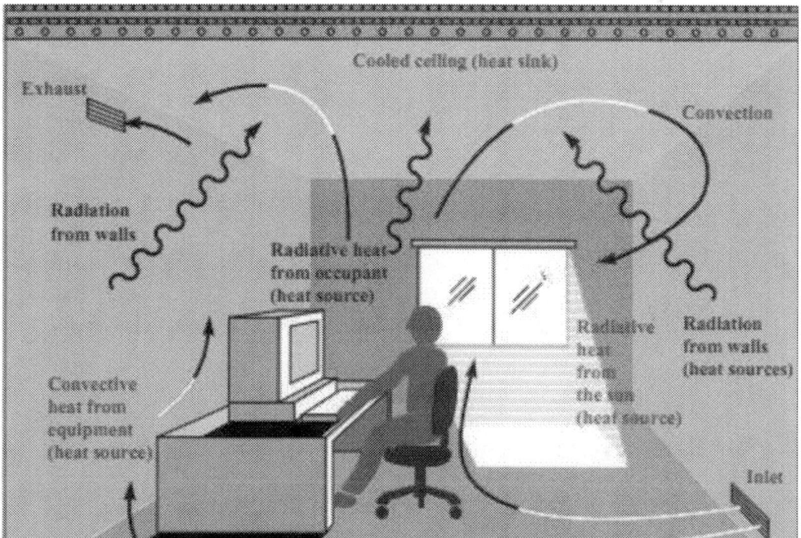

Figure 14. Radiant cooling from ceiling [43].

Dehumidification: The removal of water vapor from room air by dilution with drier air, condensation, or desiccation. In the case of condensation and desiccation, dehumidification is the exchange of latent heat in air for the sensible heat of water droplets on surfaces: both are the reverse of evaporative cooling and as such are adiabatic heating processes [39].

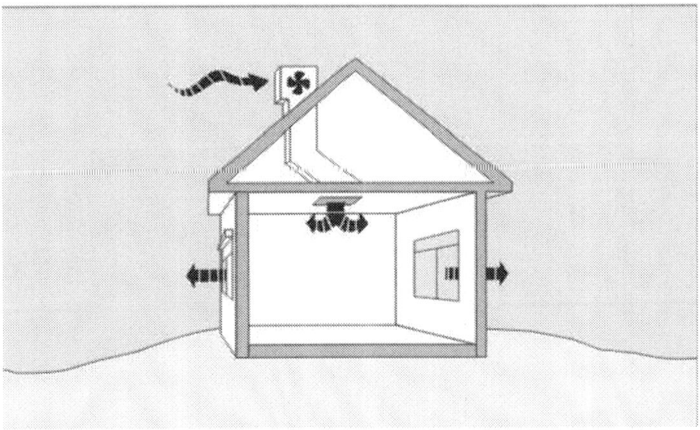

Figure 15. Evaporative cooling [40].

Mass-effect cooling: The use of thermal storage to absorb heat during the warmest part of a periodic temperature cycle and release it later during a cooler part. Night flushing (where cool night air is drawn through a building to exhaust heat stored during the day in massive floors and walls) is an example of daily-cycle mass-effect cooling, as shown in **Figure 16** [39].

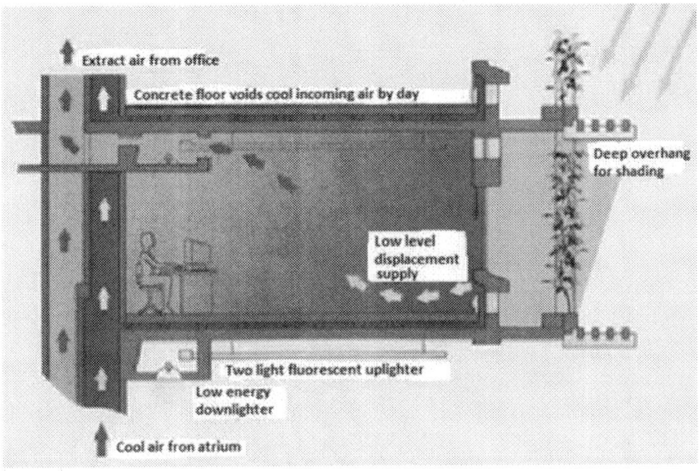

Figure 16. Air movement in Eastgate Center Office Building [44].

Usage renewable energy resources with active techniques:

The active use of solar energy systems in buildings: It is possible to produce heat and electricity with solar energy in buildings using such equipment solar collectors, photovoltaic (PV) panels, and building integrated PV (BIPV). The potential application of PV panels in high-rise buildings is more than the low-rise buildings because of higher neighboring buildings; it gives more possibility for direct solar radiation. Requirements for regulation of large amounts of PV panels are the most important problem. Because it is necessary to maintain aesthetics and PV panel's productivity in buildings[45].

The active systems where solar energy is used are the systems composed of the aggregation of mechanic and/or electronic components that convert solar radiation absorbed via collectors produced for this end into energy in a desired form and permit this to be used in building. Through these systems, solar radiation can turn into heat and electric energy [46]. These systems that transform solar radiations into energy are divided into two according to the energy they produce: solar thermal systems producing thermal energy and thermal electric (photovoltaic) systems (PV systems) producing electric energy. These systems are briefly described below.

Solar energy thermal systems: Solar energy thermal systems (effective solar thermal systems) are the aggregation of mechanic and/or electronic components that convert solar radiation into thermal energy via collectors, make it possible to directly use this energy with water, air, and a similar fluid, or make it usable by evaluating it in a storage unit. Solar energy-efficient thermal systems are used for heating pool water, preheating of climatization air and heating environment [46]. The general operation principle of thermal systems is based on collecting heat via collectors, storing thermal energy to be able to use later if needed and distributing it to relevant fields [47].

Solar water heating systems: These systems are composed of the elements that transform solar radiation into thermal energy, keep and distribute this heat in an aquatic environment. In contrary to the fact that systems show differences depending on the complexity and magnitude of necessity, all of the solar water heating systems are based on heating water, storing, and distributing it. As the hot water produced with the transformation of solar energy can be directly used for having a bath, laundry, and washing dishes depending on the characteristics of the system, it can also be used for supporting the conventional heating system [46].

Photovoltaic systems: The aggregations of the components that produce electric energy via collectors from solar radiation and make this energy usable are called photovoltaic (PV) systems. With simple or complex structuring, PV systems are used to produce electricity in a large number of different fields such as road lighting, lighthouses, vehicles, constructions, and electric power-plants. A photovoltaic system generates electric energy, stores the produced energy for necessary conditions and safely transfers this energy to the areas of usage. By being placed on fronts and roofs of buildings, photovoltaic batteries convert the solar energy coming to these surfaces into electric energy, as shown in **Figure 17** [48].

The active use of wind energy systems in buildings: Wind energy is the fastest-growing renewable energy source in the world. Wind energy is a clean fuel source and does not produce atmospheric emissions that cause acid rain or greenhouse gasses. Wind energy is an inexhaustible energy source. More recent developments in this technology have allowed wind turbines to be utilized in building design. Consistent with the high performance approach to building design, the use of wind turbines on high buildings is significantly enhanced by their integration with building architecture [50].

Figure 17. Photovoltaic panels integrated into building [49].

When the height of the structure increases the wind without interruption in direct contact with structure, wind speed increases linearly with height and utilizing the turbine at high buildings with this feature it is possible to produce significant amounts of electricity. Implementation of wind turbines in high buildings in the design stage consideration of this parameter is required: site plan layout, wind aerodynamics in building form, local wind pattern, wind speed density, frequencies of the wind speed distribution, and prevailing wind direction [51].

Must be designed taking into account the prevailing wind direction which mass form of the building and placement in the wind turbine, as shown in **Figure 18**. Previous studies show that optimal angle between the prevailing wind direction and wind turbines for maximum efficiency is determined as 45° [45].

Use of geothermal energy in buildings: Geothermal energy is used in heating and cooling in houses, greenhouse cultivation, and agriculture. Geothermal energy systems are applied in three different ways according to application methods such as heat pumps, downhole heat exchangers, and heat pipes. Their common usage in buildings is in the form of heat pipes.

Another form of geothermal energy usage is the methods where earth temperature is used. A little under earth, temperature is always in between 45 and 75 F (7.22 and 23.88°C) depending on latitude [53]. This temperature of the earth can be benefitted via air or water. The air taken through the funnels dug in various depths of earth is transferred into building and indoor is enabled to reach the same level with earth temperature. This application is ensured in the direction of heating in winter and cooling in summer. A similar application is performed to utilize the temperature of underground waters, the water circulated within the building via pipes expands the heat it has into internal volumes [48]. The schematic figure showing these applications is given in **Figure 19**.

Figure 18. Wind turbines integrated into building [52].

Use of hydrogen energy in buildings: Hydrogen energy can be used for heating houses, providing hot water, cooking and meeting electricity need. In order to use hydrogen here, we first need to produce it, then store and transfer it. Hydrogen can be produced from such renewable energy sources such as sun, hydroelectric, wind, and geothermal.

Nowadays, among the renewable energy sources, solar-hydrogen hybrid system strikes us as the most productive system. In such a system, there is a need for such constituents as photovoltaic panels, electrolyzer, fuel cell, hydrogen (H2) storing tank, battery pack, and inverter. In solar-hydrogen house energy mechanism, system works as follows [55]:

- with PV panels, electricity is produced from solar energy,

- with electrolyzer, H_2 and O_2 are produced,

- gases are taken into storage tank for heating place and water,

- in winter, by burning hydrogen "flamelessly" with catalytic hydrogen lighter (1.5 kW), the air in the ventilation system is heated,

- if additional electricity is needed, fuel cell runs,

- a part of the heat emerging in fuel cell is used to heat water.

Use of biomass energy in buildings: Biomass is a strategical energy resource, which is renewable and environment friendly, can be grown everywhere, enables socioeconomic improvement, and can be used for power generation and for obtaining fuel for vehicles. Biomass is utilized in the energy sector by being directly burned or its fuel quality is increased with various processes and thus gained alternative biofuels (easily movable, storable, and usable fuels), which are equal to existing fuels.

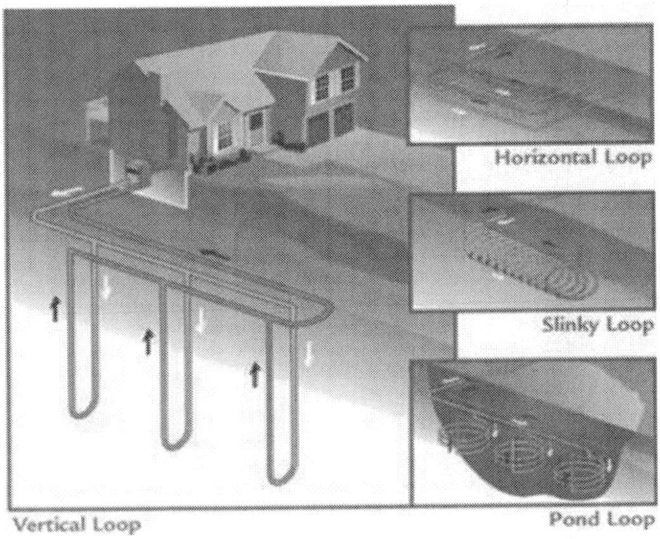

Figure 19. Earth-based heat pipe usage forms [54].

From biomass is produced fuel with physical processes (size reducing-breaking and grinding, drying, filtration, extraction and briquetting) and transformation processes (biomassive and thermochemical processes) [56]. In the houses, biomass is used for biogas power generation adopted with airless increpation, ethanol heating adopted with the pyrolysis method, and hydrogen water heating adopted with the directly burning method [47].

Natural lighting: Natural lighting in buildings is carried out through a most basic windows and skylights. Choice of direction in the windows and roof lighting is important. The most suitable directions for natural lighting are south and north. The north direction is not exposed to radiation, but can always get daylight in the same quality. In the west and east directions, the sun radiates in horizontally and makes it difficult to control. In the south direction, the effect of the sun is permanent and sun rises at a right angle compared to the west and east directions. Therefore, it is easy to control [57].

http://dx.doi.org/10.5772/66670

In order to increase daylight's entrance into the building, light colors should be used on windows' wings that direct the light and light shelves. Moreover, the elements used to reflect light must be made in a position to reflect the light to the ceiling. Wall and ceiling surfaces must be light colored so that the light can be spread [58]. Desirable reflectance according to Illuminating Engineering Society's recommendations: ceilings >80%; walls 50–70% (higher if wall contains window); floors 20–40%; and furniture 25–45% [58, 59].

The proper design and selection of the daylighting systems can help in improving energy efficiency and reducing environmental pollution. Windows, clerestories, and roof monitors when properly designed can provide of the lighting needs without undesirable heat gain and glare. And therefore, electric lights can be turned off or dimmed in day-lit spaces when the target illuminance is achieved by daylighting. Energy savings can only be achieved by implementing light controls, sensors, and light dimmers for the lighting system of those day-lit spaces. The usage of daylight in buildings decreases the electric energy consumption. For instance, it has been shown that artificial lighting of nondomestic buildings represents 50% of the energy consumption in Europe. It also has been shown that it is possible to reduce this consumption by between 30 and 70% by combining the use of artificial and natural lighting. Potential savings depend on orientation, the size and shape of the window, and the shape and surface reflectance of the room [60–64].

Another usage of natural lighting in buildings is the use of the daylighting system. Daylighting is defined as "the combination of the diffused light from the sky and sunlight." A daylighting system preferred function is to redirect a significant part of the incoming natural light flux to improve interior lighting conditions, therefore located near or in the openings of building envelope. Daylighting systems are divided into two categories: side-lighting and top-lighting. Light can come from many types of glazing configurations, which are either vertical or horizontal and from the side or from the top. Side-lighting, which is more commonly observed, is simply a window opening. Top-lighting is an opening in the ceiling or roof element of the building [60]. Applications of the daylight system are discussed below, as shown in **Figure 20**.

2.2. Energy-efficient designing methods in building phase

Building phase includes the construction and usage processes of building. Building phase is possible with preferring building techniques consuming less energy and using energy-efficient equipment. The energy used in construction changes according to building systems. For instance, in a study carried out by Hozatlı and Günerhan, it was established that frame construction consumes less energy than a reinforced concrete frame construction during its life cycle [66]. As the energy consumption of the buildings constructed with different materials changes, energy consumption also changes in the buildings constructed with the same materials. The energy consumption of the commonly used reinforced concrete frame building system was analyzed according to three different building methods as follows:

1. Conventional frame building system: The most prominent characteristic of the conventional building system is that the whole of production is implemented in building site thanks to intensive man power. When it is analyzed with respect to energy consumption, the energy consumption of the conventional system is at a low level due to the characteristics of the equipment (concrete mixer, roof crane) used in the stages of concrete production and concrete casting [66].

2. Tunnel form concrete masonry system: Tunnel form masonry system requires a certain preliminary investment. The system is suitable for large scale and permanent productions. Because lifting cranes consuming a lot of energy are used to carry big and heavy forms, energy consumption is high. The task of curing with concrete plant and intratunnel heaters raises the energy consumption of the system [67].

3. Precast construction systems: Since the majority of the processes realized in the building area in other systems are made in the manufacturing plant, energy consumption is very high. In these systems, downloading components from transportation vehicles to worksite, their storage and mounting are performed by lifting cranes. For this reason, at these stages as well, the high amount of energy is consumed. While heavy duty vehicles transporting ready building elements from manufacturing plant to building site lead to problems in traffic, they also increase the consumption of energy [67].

As a result, it can be urged that in a tunnel form, precast framework and precast panel building systems, manufacturing processes create negativities to a large extent in terms of energy consumption and emissions, energy consumption in the conventional systems is quite less compared with these but more negative with respect to the formation of solid waste [67].

As it is understood from the reinforced concrete building system, there are different methods for the same material in the construction of buildings. Such heavy duty vehicles as lifting cranes, concrete pumps, and concrete transit mixers consume the high amount of energy. For this reason, the building methods consuming less energy should be preferred on the condition that no concession is made in the quality of building.

Usage process is the process that consumes most energy in buildings. According to WBCSD (World Business Council for Sustainable Development) report, 88% of the energy consumed in buildings is spent during usage and maintenance [68]. The applications mentioned in designing process gain buildings energy efficiency during usage period. Besides this, the applications below have also the potential of procuring considerable energy saving during usage period.

Supporting multiuse improvement: Sustainable development advocates the combination of house settlement, trading area, office, and retail areas. Thus, people get the opportunity of living next to the places they work and shop. This renders the formation of a community different from traditional suburbs. 24-hour activity potential also makes the land safer [68].

Combining design with public transportation: Sustainable architecture on urban scale should be designed in a way to support public transportation. Thousands of vehicles coming in and going out the land during daily work pressure cause air pollution and traffic jam and they need parking areas [69].

Using energy-efficient bulbs and energy-efficient appliance: For example, the light-emitting diode (LED) is one of today's most energy-efficient and rapidly developing lighting technologies.

Lighting controls: Lighting requirements reply to a building design. The need for lighting, when during daytime, will depend on the window size and placement, and the position of buildings. The need for lighting is decreased by the use of automatic controls, which depend on the orientation of building windows, the supply of daylight, and usage of the room [70].

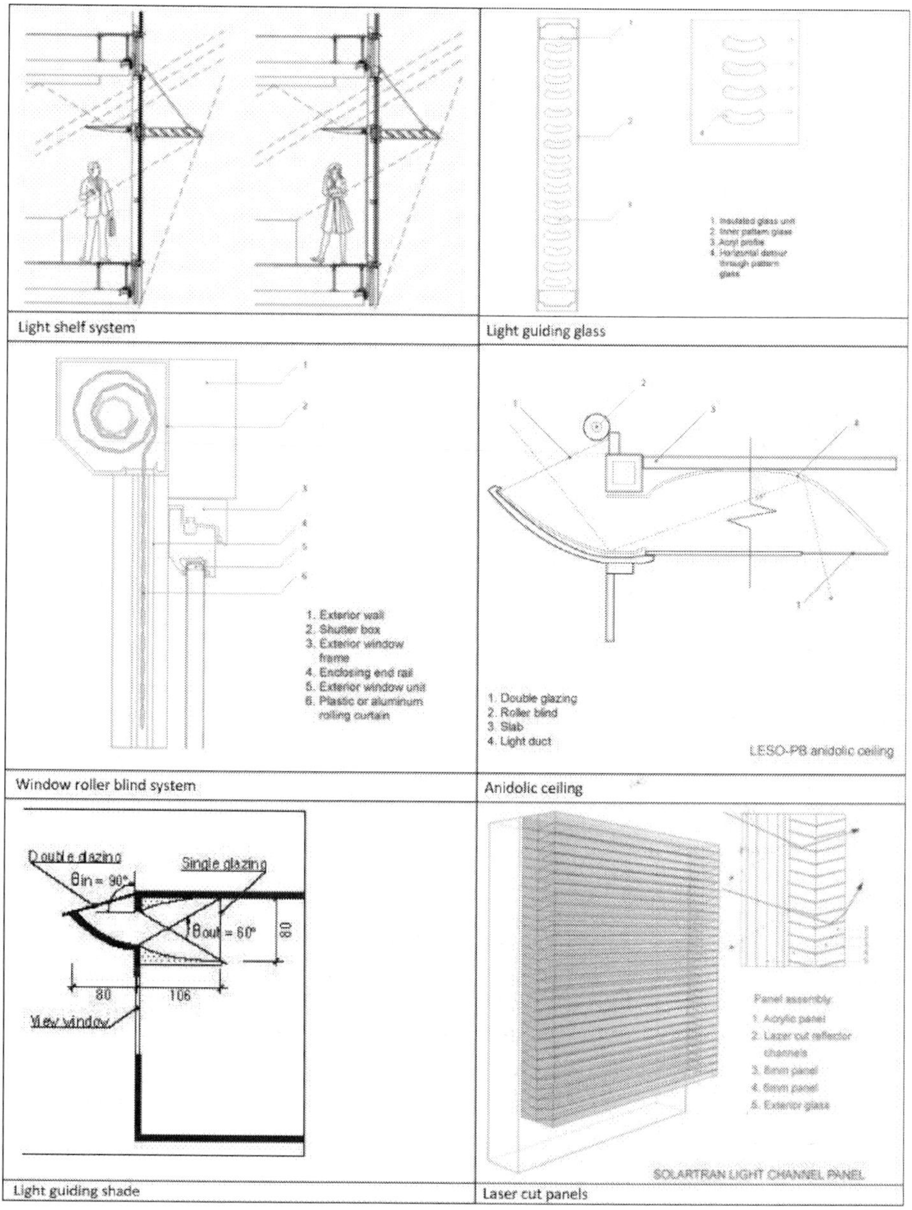

Figure 20. Daylighting systems using direct sunlight [57, 65].

High-efficiency heating, ventilation, and cooling equipment: Heating ventilating air conditioning (HVAC) systems extremely influence energy consumption in buildings. The relationship between building specifications and HVAC systems are: highly efficient building envelopes reduce the need for heating and cooling systems. Good and intelligent designed buildings can reduce the need for HVAC systems. Efficiency improvements in HVAC systems can lead to substantial savings. If, for instance, energy efficiency is improved in a heating boiler or an air-conditioner, total savings will depend on the total need for heating or cooling in the building. In a well-insulated building envelope, the energy needs of the HVAC system are reduced. The building can be separated into thermal zones at suitable dimensions, reducing the need for heating, cooling and ventilation with careful building planning [70].

2.3. Energy-efficient methods in postbuilding phase

The postbuilding phase is the phase when the usage phase is completed. This phase includes the demolition of building, recycling, and destruction of it. In this phase, it is important to recycle the building materials and compositions used in the buildings and reuse buildings. After the functional uses of buildings have been completed, reutilization of them in other functions instead of demolishing them protects such resources as raw material, water, and energy. It should be enabled to reuse the building compositions of the buildings, for which demolition decision has been taken, such as the roof truss, woodworks. Following the saving of appropriate building compositions, recyclable building materials are needed to be separated. In this way, raw material protection is provided for the building material to be reproduced, and thus it is saved from the energy to be consumed while processing raw material.

It is necessary to use machines and equipment as few as possible while demolishing buildings and select equipment procuring energy saving for demolition.

3. Conclusion

Buildings have a huge potential for energy efficiency. To obtain these large potential there is a need to take some regulations and initiatives to improve the efficiency in buildings. Energy consumption in buildings occurs in every phase of building life cycle. However, the important phase is the usage and maintenance process of buildings where energy is consumed most within the scope of life cycle. During the building life cycle, the highest energy consumption occurs during the usage stage. This is because this period is much longer in duration compared to the other stages and the comfort levels necessary for human health and working efficiency need to be provided at this stage. Therefore, in energy-efficient building designs, especially the usage stage should be taken into consideration. In order to reduce energy consumption in the usage process of building, renewable energy sources instead of fossil-based energy sources should be preferred. Importance to the use of renewable energy sources should be given. In particular, the use of active and passive systems should be noted. Energy simulation programs in building design should be used.

Within the scope of this study, energy-efficient strategies described under the titles of the choice of settlement in the prebuilding phase, planning settlement, building form, building organization and planning, building envelope, choice of building material, landscape design, and utilization of renewable energy sources are directed to using less energy or cleaner energy in the usage process of buildings. Examining these strategies, it is clear that cooperation of very different disciplines (architecture, mechanical engineering, civil engineering, landscape architecture, urban and regional planning and interior architecture) is required. For this reason, designing an energy-efficient building is possible with a multidisciplinary study which begins from the emergence of the idea of constructing a building and lasts until the demolition of the building at the end of usage period.

Author details

Izzet Yüksek[1]* and Tülay Tikansak Karadayi[2]

*Address all correspondence to: izzetyuksek@gmail.com

1 Fine Arts Design and Architecture Faculty, Celal Bayar University, Manisa, Turkey

2 Architecture Faculty, Gebze Technical University, Gebze, Kocaeli, Turkey

References

[1] Briefing note 33: The Impacts of Construction and the Built Environment, Willmott Dixon Re-Thinking Ltd. (2010). URL: www.willmottdixon.co.uk/asset/download/9462

[2] Lenzen M, Treloar GJ, Embodied energy in buildings: wood versus concrete-reply to Börjesson and Gustavsson, Energy Policy, 2002; 30:249–244. DOI: 10.1016/S0301-4215(01)00142-2

[3] Lam JC, Wan KKW, Tsang CL, Yang L, Building energy efficiency in different climates, Energy Conversion and Management, 2008; 49(8):2354–2366. DOI: 10.1016/j.enconman.2008.01.013.

[4] Anonymous, Lighting the Way: Toward a Sustainable Energy Future, Inter Academy Council. (2007). URL: http://www.interacademycouncil.net/File.aspx?id=24548

[5] Mohsen MS, Akash BA, Some prospects of energy savings in buildings Amman, Energy Conversion and Management, 2001;42(11):1307–1315. DOI 10.1016/S0196-8904(00)00140-0.

[6] Lam JC, Wan KKW, Tsang CL, Yang L, Building energy efficiency in different climates, Energy Conversion and Management, 2008;49(8):2354–2366. DOI: 10.1016/j.enconman.2008.01.013

[7] Utkutuğ G, Yeşil Mimarlık (Green Architecture), Bilim Teknik Dergisi, Yeni Ufuklara

Eki, 2002, 420, Ankara

[8] Zeybek Ö, Energy Savings of Primary School Building in Turkey Through Energy Efficient Renovation Strategies in Building Envelope, [thesis]. Istanbul: İstanbul Technical University, Institute of Science and Technology; 2009.

[9] Özdemir BB, Sürdürülebilir Çevre İçin Binaların Enerji Etkin Pasif Sistemler Olarak Tasarlanması, (Energy efficient building design as passive systems for a sustainable environment) [thesis]. Istanbul: Istanbul Technical University, Graduate School of Engineering and Sciences; 2005.

[10] Yılmaz AZ, Akıllı Binalar ve Yenilenebilir Enerji (Smart Buildings and Renewable Energy), 7. Ulusal Tesisat Mühendisliği Konferansı, 23–26 Kasım 2005, İzmir, pp. 1–8.

[11] Olgyay V, Design with Climate Bioclimatic Approach to Architectural Regionalism, Princeton, New Jersey: Princeton University Press; 1969.

[12] Lechner N, Heating, Cooling, Lighting Design Methods for Architects, Canada: John Wiley Sons; 1991.

[13] Berköz E, Küçükdoğu M, Yılmaz et al. Enerji Etkin Konut ve Yerleşme Tasarımı (Energy Efficient Housing and Settlement Design), TÜBİTAK, INTAG 201, İstanbul; 1995.

[14] Soysal S, Konut Binalarında Tasarım Parametreleri ile Enerji Tüketimi İlişkisi (Relation Between Design Parameters and Energy Consumption in Housing Buildings), [thesis]. Gazi University, Ankara; 2008..

[15] Karagözlü BA, Konutlarda Enerji Giderlerinin Azaltılmasına Yönelik Bir Çalışma, [thesis]. Istanbul: İstanbul Technical University; 2006.

[16] Arınç ÜD, Enerji Verimli Binaların Yerleşim ve Tasarım Esasları (Principles of Settlement and Design of Energy Efficient Buildings) , Akıllı ve Yeşil Binalar Kongresi, 23–24 Mayıs 2013, Ankara, pp. 19–29.

[17] Chance T, Towards sustainable residential communities; the Beddington Zero Energy Development (BedZED) and beyond, Environment & Urbanization, 2009; 21(2): 527–544.

[18] Göksal T, Özbalta N, Enerji Korunumunda Düşük Enerjili Bina Tasarımları (Low Energy Building Designs in Energy Conservation), Mühendis ve Makine, 2002; 43(506):26–32.

[19] Daniels K, The Technology of Ecological Building: Basic Principles and Measures, Examples and Ideas, Basel Boston: Birkhauser Verlag; 1979.

[20] Yüksek İ, Esin T Analysis of traditional rural houses in Turkey in terms of energy efficiency, International Journal of Sustainable Energy, 2013; 32(6):643–658.

[21] Nikolic V, Bau und Energie, Bauliche Mabnahmen zur verstarlen Sonnenenergienutzung im Wohnungsbau. Herausgeber: Der Bundesminister für Forschung und Technolgie, Verlag TÜV. Rheinland, Köln, Destschland; 1983.

[22] Yüksek İ, A Study on Ecological Assessment in Traditional Anatolian Architecture (A Case Study: Rural Area of Kırklareli Province), [thesis]. Trakya University; 2008.

[23] Lorenz P, Bauen+Wohnen Umweltbewusst und Kostensparend (Building + Living Environment-conscious and cost-saving), Pietsch Verlag, Stutgart; 1988.

[24] Olgyay V, Design with Climate Bioclimatic Approach to Architectural Regionalism, Princeton, New Jersey: Princeton University Press; 1969.

[25] Çakmanus İ, Böke A, Binaların Güneş Enerjisi İle Pasif Isıtılması ve Soğutulması (Passive Heating and Cooling of Buildings with Solar Energy), Yapı Dergisi. 2001; 235:83–88.

[26] Donnelly J, Energy Efficiency in Traditional Buildings, Dublin, Ireland: Stationery Office, 2011. http://www.dublincity.ie/Planning/HeritageConservation/Conservation/Documents/Energy%20Efficiency%20in%20Traditional%20Buildings.pdf

[27] Naidj JS, A Comparative Study of Passive Solar Building Simulation Using Hot2000, TRN-SYS14, NETSPEC, [thesis]. Trent University, Peterborough, Ontario, Canada; 1998.

[28] Dirim TV, Comparison of Different Settlement Arrangements of a Mass Housing Project in Istanbul in Terms of Energy Efficiency, [thesis]. Yıldız Technical University; 2014.

[29] Demirbilek N, Eryıldız DI, Güneş Mimarlığı (Solar Architecture), Temiz Enerji vakfı Yayınları, Ankara 2001.

[30] Esin T, Appropriate material selection for sustainable building, Building Magazine, 2006; 291:83–86.

[31] Berge B, The Ecology of Building Materials, Oxford: Architectural Press; 2009.

[32] Huberman N, Pearlmutter AD, Life cycle energy analysis of building materials in the Negev desert, Energy and Buildings, 2008; 40(5):837–848. DOI: 10.1016/j.enbuild.2007.06.002

[33] Yüksek İ, The Evaluation of Building Materials in Terms of Energy Efficiency, Periodica Polytechnica-Civil Engineering, 2015; 59(1):45–58. DOI: 10.3311/PPci.7050

[34] Esin T, Yapılarda Etkin Enerji Kullanımı – Sürdürülebilir Yapılaşma İçin Öneriler (Effective Energy Use in Buildings - Proposals for Sustainable Convergence), Kritik Teknolojiler Sempozyumu, 20–21 September 2001; Kocaeli, Turkey, pp. 393–404.

[35] Haque MT, Tai L, Ham D, Landscaping design for energy efficiency, South Carolina Energy Office, Clemson University Digital Press, USA, 2014.

[36] Low-Energy Building Design Guidelines, Energy-efficient design for new Federal facilities, Federal Energy Management Program Prepared by the New Technology Demonstration Program. Prepared for the U.S. Department of Energy by the National Renewable Energy Laboratory, a DOE National Laboratory DOE/EE-0249 July 2001.

[37] USAF, Passive Solar Handbook, Introduction to Passive Solar Concepts, Volume I, United States Air Force, Colorado, USA, 2015.

[38] Cook J, Passive Cooling, Cambridge, Massachusetts: MIT Press, 1990; pp. 1–42.

[39] Moore F, Environmental Control Systems, Heating Cooling Lighting, McGraw - Hill, Inc. New York, USA, 1993.

[40] Cooling Your Home with Fans and Ventilation, DOE/GO-102001-1278, FS228, June 2001, NREL, US Department of Energy. URL: http://www.nrel.gov/docs/fy01osti/29513.pdf

[41] http://www.cce.ufl.edu/current/green_building/passive.html [Accessed: 2016-04-05]

[42] http://www.dyerenvironmental.co.uk/natural_vent_systems.htm [Accessed: 2016-05-07]

[43] http://www.oorja.in/what-is-radiant-cooling/ [Accessed: 2016-05-07]

[44] Slessor, C, Critical mass - use of principles of natural heating, cooling and ventilation for building in Harare, Zimbabwe. Architectural Review, The. FindArticles.com. 16 Jan, 2011. http://findarticles.com/p/articles/mi_m3575/is_n1195_v200/ai_19007171/

[45] Sev A, Başarır B, Geçmişten Günümüze Enerji Etkin Yüksek Yapılar ve Uygulama Örnekleri (Sustainable Energy-efficient High-D Construction and Application Examples from the Past), X. Ulusal Tesisat Mühendisliği Kongresi, 13–16 Nisan 2011, İzmir, Türkiye, pp. 1499–1513.

[46] Sakınç E, Sürdürülebilirlik Bağlamında Mimaride Güneş Enerjili Etken Sistemlerin Tasarım Öğesi Olarak Değerlendirilmesine Yönelik Bir Yaklaşım (An approach to evaluate solar active systems as an architectural design criteria in the context of sustainability) [thesis]. Yıldız Technical University; 2006.

[47] Şerefhanoğlu M, Güneş Işınımlarından Yararlanma ve Korunma (Use and Protection of Sunlight) YTÜ Basımevi, İstanbul; 1988.

[48] Yüksek İ, Esin T, Usage Opportunities of Renewable Energy Resource in Buildings, 5th International Advanced Technologies Symposium (IATS'09), May 13–15, 2009, Karabuk, Turkey, pp. 1848–1853.

[49] http://www.baulinks.com/webplugin/2005/i/1022- rehau.jpg [Accessed: 2008-11-11]

[50] Advantages and challenges of wind energy, http://energy.gov/eere/wind/advantages-and-challenges-wind-energy, [Accessed: 2016-05-07]

[51] Frechette, R. Gilcgrist, R., Towards Zero Energy: A Case Study of Pearl River Tower, Guangzhou, China. CTBUH 8th World Congress, 2008 March 3–5, Dubai.

[52] http://farm3.static.flickr.com/2359/2139274930_5c7be 4866d.jpg [Accessed: 2008-11-11]

[53] [http://www.earthenergysystems.com/for_architects [Accessed: 2008-11-11]

[54] http://www.kcpl.com/residential/images/geoheatpump.jpg [Accessed: 2008-11-11]

[55] Tabakoğlu Ö, Hidrojen Enerjisi & Hidrojenin Binalarda Kullanımı (Hydrogen Energy &

Use of Hydrogen in Buildings), Enerji Ekoloji Paneli, Diyarbakır; 2007.

[56] Karaosmanoğlu F, Yenilenebilir Enerji Kaynakları ve Türkiye (Renewable Energy Research and Turkey), Görüş Dergisi, İstanbul, Mart-Nisan; 2003.

[57] Yüksek İ,Görgülü S, Kocabey S,Tuna M, Dursun B. Assessment of daylighting performances of the classrooms: a case study in Kirklareli University, Turkey. Light & Engineering, 2015; 23(1):32–38.

[58] O'Connor J, Lee E, Rubinstein F, Selkowitz S, Tips for daylighting with windows, Berkeley: Building Technologies Program. Energy & Environment Division, Ernest Orlando Lawrence Berkeley, National Laboratory, University of California, 1997.

[59] Illuminating Engineering Society of North America, IESNA lighting handbook, Reference and Application 9th edition. New York, IESNA, 2000.

[60] Alrubaih MS, Zain MM, Alghoul MA, Ibrahim NN, Shameri MA, Elayeb O, Research and Development on Aspects of Daylighting Fundamentals. Renewable and Sustainable Energy Reviews, 2013; 21:494–505.

[61] Tetri E, Daylight linked dimming: effect on fluorescent lamp performance. Lighting Research and Technology, 2002; 34(1):3–10.

[62] Mjos T, Larsen P, Indoor lighting–energy friendly installations, Light & Engineering, 2012; 20(2):27–38.

[63] Abdelatia B, Marenne C, Semidor C, Daylighting strategy for sustainable schools: case study of prototype classrooms in Libya. Journal of Sustainable Development, 2010;3(3):60–67.

[64] Cantin F, Dubois MC, Daylighting metrics based on illuminance, distribution, glare and directivity. Lighting Research and Technology, 2011; 43(3):291–307.

[65] http://d-lite.org/ [Accessed: 2016-05-24]

[66] Hozatlı B, Günerhan H, Life cycle assessment of buildings constructed with reinforced concrete and wood material in terms of Muğla province, Engineer and Machine, 2015;56(660):52–60.

[67] Esin T, Çoşgun N, Betonarme Yapım Sistemlerinin Ekolojik Açıdan Değerlendirilmesi (Ecological Evaluation of Reinforced Concrete Construction Systems), 2. Ulusal Yapı Malzemesi Kongresi ve Sergisi 6–8 Ekim 2004, İstanbul.

[68] World Business Council for Sustainable Development, Energy Efficiency in Buildings, Business realities and opportunities, Full Report, Geneva; 2007.

[69] Çelebi GÜ, Gültekin AB, Harputlugil GU, Bedir M, Tereci A, Yapı-Çevre İlişkileri (Building-Environment Relations) TMMOB Mimarlar Odası, Sürekli Mesleki Gelişim Merkezi Yayınları-10, İstanbul; 2008.

[70] Laustsen J, IEA (International Energy Agency) Energy Efficiency Requirements in Building Codes, Energy Efficiency Policies for New Buildings IEA Information paper, International Energy Agency © OECD/IEA, 2008.

Section 2

Building Energy Consumption, Demand and Efficiency

Chapter 6

Energy Efficiency in Manufacturing Facilities: Assessment, Analysis and Implementation

Gary P. Moynihan and Frank L. Barringer

Additional information is available at the end of the chapter

http://dx.doi.org/10.5772/64902

Abstract

Manufacturing facilities are one among the largest consumers of energy. Efforts to improve energy efficiency are an increasing concern for many manufacturing facility engineering managers. This can be achieved by evaluating energy end uses (e.g., lighting, processing equipment, and heating, air conditioning, and ventilation (HVAC) systems), and by implementing measures to reduce the total amount of energy consumed for one or more of the end uses. Throughout the 40 years of its existence, the US Department of Energy's Industrial Assessment Center program has developed an array of techniques to improve energy efficiency in industrial facilities. This chapter discusses recommended assessment procedures and observed energy-saving opportunities for some of the most common industrial energy systems. These systems include lighting, compressed air, boilers and steam systems, manufacturing process heating, HVAC, pumps, and fans. Implementation of these assessment recommendations has been demonstrated to increase efficiency and to thus reduce energy consumption and associated costs. While every manufacturing facility is different, and their component industrial energy systems equally unique, this chapter presents a set of analytical guidelines that can be used as a template by engineering practitioners to evaluate their facility energy use and implement subsequent energy conservation measures.

Keywords: energy management, energy conservation, manufacturing, industrial facility, efficiency

1. Introduction

As noted by Vanek and Albright [1], manufacturing facilities are one of the largest consumers of energy in the United States, comprising approximately 32% of the energy end use in the country. Efforts to improve energy efficiency are an increasing concern for many facility engineering managers. Improved use of energy in a building provides lower operating costs. This can be achieved by evaluating energy end uses (e.g., lighting, processing equipment, and heating, air conditioning, and ventilation (HVAC) systems), and by implementing measures to decrease the total amount of energy consumed for one or more of these end-use systems.

The United States Federal Department of Energy (DOE) has developed a number of programs to address energy conservation and energy management in the US industrial sector. The Industrial Technologies Program (ITP), of the Energy Efficiency and Renewable Energy (EERE) division of the DOE, has a series of initiatives to assist manufacturers in identifying areas where energy use may be decreased, as well as indicating those areas where excess energy could be redirected for other purposes [2].

One such program regards the establishment of Industrial Assessment Centers (IACs). The goal of these centers is to increase energy efficiency for small-to-medium-sized manufacturing companies throughout the United States. The focus of their recommendations is for building energy retrofits that help lower the amount of energy used in the manufacturing plant. IAC teams consist of students and faculty, from the universities participating in this effort, who conduct these energy audits and develop the subsequent energy-saving recommendations [2].

The IAC's energy audit of a manufacturing facility is conducted in three phases, consistent with the procedure detailed by Thurman and Younger [3]. The first phase involves analyzing the data from the facility energy bills to determine what energy is used and how this energy use varies over time. The second phase consists of a factory walk-through inspection and audit, looking carefully at each end-use system within the facility and recording information for later use [3]. The last phase is where specific energy savings are identified for later implementation by the facility. The IAC team performs a detailed analysis supporting specific recommendations with related estimates of costs, performance, and energy savings.

2. Lighting systems

In 2010, lighting consumed 700 TWh (terawatt hour) of site energy, which is 18% of the total electricity consumption of the United States. Of the 700 TWh of electricity, industrial lighting consumed 26% [4]. There are different lamps and lamp fixture. High-intensity discharge lights include metal halide, high-pressure sodium, low-pressure sodium, and mercury vapor, while fluorescent lights include T12, T8, and T5 linear bulbs and compact fluorescent lamps [5, 6]. High-intensity discharge (HID) and fluorescent lights are the most common lights in industrial facilities. Other lamps include light-emitting diode, induction, and incandescent lighting [7]. The most common types of fluorescent lamps are T12, T8, T5, and compact fluorescent (CFL). The number after the "T" corresponds to the diameter of the lamp in eighths of an inch (Tx equals x/8 inches in diameter). The T12 lamp is a linear tube that in twelve-eighths inches in diameter, or 1.5 inches in diameter. The T8 lamp is 1 inch in diameter, while the T5 is five-eighths inches in diameter. CFLs were created to replace incandescent and HID lamps, and consist of an endless arc pathway formed by multiple tubes joined together [6]. Fluorescent lamps are also required to operate in series with a current-regulating device or ballast.

A number of controls are used to operate lighting systems; these controls include occupancy sensors, photosensors, switches, and automatic control systems that can control lighting systems along with other systems in a facility. Depending on the manufacturing process or task that is being performed, the light level requirements can range from five to 500 foot-candles [8]. The Illuminating Engineer Society [8] provides guidance on these industrial-lighting level requirements.

High-intensity discharge and fluorescent lights require a ballast to start and operate the lights [6]. These ballasts operate with a ballast factor or ballast efficiency that is normally greater than one. Ballasts limit the current to the correct amount that a lamp is designed for, and ballasts deliver the essential starting and operating voltages to lamps. High-intensity discharge lights use constant-wattage autotransformers, constant-wattage ballast, lag (reactor) ballasts, magnetic regulator (constant-wattage) ballasts, and lead circuit ballasts for operation. For fluorescent lights, magnetic and electronic are the two types of ballasts that are used for operation [5].

2.1. Recommended assessment procedures

After the assessment team arrives at the facility, a kick-off meeting with the team members and facility personnel will take place to explain the purpose of the assessment, the process of the assessment, and the timeline for completion [3]. After the kick-off meeting, a tour of the lighting systems will take place. Once the tour is finished, data collection may begin. Each different lighting system needs to be identified, and the each lamp in the lighting system needs to be counted and recorded on a data sheet [3]. Once the type of lamp is observed, the wattage of each lamp and the ballast factor of each fixture need to be recorded. The plant personnel should be able to assist if questions arise. Alternatively, replacement lamps and ballasts may be found in the maintenance parts storage, and may provide this information. Further, a ballast detector can be used to determine the type of ballast in each light, that is, whether it is electronic or magnetic.

The operating hours of each lighting system need to be recorded and this information should come from the facility personnel. If different areas in the plant have different lighting systems, then the lighting system and type of lights that are in those areas should be recorded while counting the lights. In each different area the amount of personnel traffic should be recorded and along with the amount of vacant time. The controls and schedule of each lighting system should be recorded and this information should be given by the plant personnel. Before the lighting levels are taken using the light level meter, the daylighting panels or windows should be noted [4]. Finally, the light levels in each area need to be measured and recorded. Several measurements should be taken per area, especially in areas with critical lighting requirements. The activities and processes in each area need to be recorded, to be able to make sure the correct amount of light for the activity is reached at the work plane.

The number of each type of light should be gathered together along with the wattage, ballast factor, and operating hours. After that, these four numbers can be multiplied together to get the annual energy usage of the lighting systems [5].

2.2. Recommended assessment opportunities

One of the most common lighting system assessment recommendations (ARs) is replacing HID lighting with more efficient lighting. Either linear fluorescent tube T8 or T5 lighting or CFLs are good replacements. These lights normally have a better lumen maintenance for better light as the lamps age, and normally have a higher color rendition index (CRI) for a better differentiation in color of light (particularly needed for many quality inspection tasks) [5].

Another lighting system AR is replacing T12 fluorescent lamps with T8 fluorescent lamps. Replacing T12 lamps with T8 lamps produces an opportunity for energy and cost savings. T8 lamps have less wattage, but produce more lumens than T12 lamps [9]. The T8's life hours and ballast factor can be the same, higher, or lower than the T12's.

Incandescent lamps may also be replaced with CFLs. Replacing the incandescent lamps with CFL lamps produces an opportunity for energy and cost savings by reducing the wattage for the same amount of lumens as an incandescent. For example, 200-watt CFL lamps have less wattage, but produce more lumens than 500-watt incandescent lamps [9]. Further, the CFL has a longer expected life than the incandescent lamp.

Installing occupancy sensors (motion sensors) is a possible lighting system AR for consideration. Occupancy sensors will turn off the lights when an area is vacant. The lights should stay on a minimum of 30 min after they turn on, so as long as there are gaps in the area's traffic longer than 30 min energy savings will be seen. The only drawback to an occupancy sensor AR is that the lights have to be rapid start lighting systems or they will not work correctly [5]. Installing occupancy sensors in certain areas of the plant can result in a large energy and cost savings by reducing the operating hours of a lighting system.

A related AR is incorporating photosensors with lights in areas where adequate daylighting is available during daytime hours or in areas where daylighting panels can be installed. Photosensors will turn off the lights when the daylighting is adequate to provide light to certain areas of the plant. An indication of this possible AR is if areas around the perimeter of the plant have adequate daylighting available to help with illuminating the plant floor or warehouse [3]. By discovering daylighting panels in the plant, and measuring the light levels in the areas that have these panels, the photosensors AR can be decided upon. The light levels need to be compared to the light level requirements for the specific processes in the area. By installing photosensors in areas around the perimeter of the plant where daylighting is adequate for the lights to be shut off, then this AR will result in energy and cost savings.

De-lamping (reducing the quantity of lights) in the facility is a possible energy-saving AR. Evidence for the need for this AR is observation of areas of the plant where light levels exceed the requirements of that area and/or have high bay lights that can be lowered to increase light levels without hindering the processes [3]. Conducting light level measurements and looking at the heights of lights will help determine if there are too many lumens in a certain area or if the lights can be lowered [10]. De-lamping areas of the facility will reduce the energy for the lighting system by using the actual required amount of light for each area [9]. Lowering the lights down from the ceiling can also make de-lamping available by having more lumens at the work plane.

Turning off the lights when areas of the facility or the entire facility are vacant will save energy for the lighting systems by reducing the operating hours of the lights. This AR will also allow the lights to last longer and reduce replacement costs. The energy savings can be calculated by reducing the operating hours of the lighting system [10].

3. Compressed air systems

Nearly all industrial facilities have compressed air systems, and most could not operate without it. Inefficiencies in compressed air systems can be very significant, and energy-saving projects can range from 20 to 50% of electricity consumption [11]. Compressed air systems are categorized as supply side and demand side. The supply side of a compressed air system consists of a compressor, the prime mover, control system, air dryer, air filter, and storage. The compressor, prime mover (motor), and control system are all contained in the unit's package. The compressor package also contains a cooling system, which can be either air-cooled or water-cooled. The air that is discharged from the compressor will normally flow through an air dryer and air filter before going to air storage prior to the demand side. The demand side of a compressed air system encompasses the piping distribution system, dedicated air receivers, pressure/flow controllers, filters, regulators, lubricators, and end uses. Normally, the air receiver is located close to the compressors, or immediately before the end use, and the filters, lubricators, and regulators are also in close proximity to the end uses [11].

3.1. Recommended assessment procedures

After the kick-off meeting, the team should install power and airflow data collection equipment and also retrieve power and airflow data from the data systems that are permanently installed into the compressed air systems [3]. While the data are being collected or before the end of the assessment visit, the collected data should be validated to assure all the data are accurate and accounted for. Production process operating data and plant functions should be gathered to establish a functional baseline for the plant [11]. While conducting the assessment, a comprehensive plan to observe and measure the supply-side performance of the compressed air system should be completed. Once the supply side is observed and measured, then the transmission from the supply side to the demand side should be observed and any required measurements should be taken [10]. Finally, the end-use applications (demand side) should be observed and measured.

The nameplate data are located on the side of the compressor and should be recorded in the assessment notes, and a picture should be taken of the nameplate if available. The "cut in" and "cut out" pressure should be recorded in the notes, and these pressure values can be found on the compressor's display screen. The air pressure upstream and downstream of the cleanup equipment (i.e., filters, dryers, etc.) should be measured with a reliable gauge [10]. The operating pressure is assumed to be the midpoint between the cut in and cut out pressures or this pressure can be measured [12]. The plant-floor pressures should be measured with a reliable gauge at many different places on the plant floor. Next, the controls of the compressor should be identified from the control panel or from plant personnel. Plant personnel should also be able to explain how the schedule of the compressed air system is established, since the operating hours along with the non-operating hours are critical to assessing the energy usage. Knowing how the schedule is controlled is also critical, that is, whether the schedule is controlled by plant personnel or a sequencer [10].

The cooling method for the compressors needs to be recorded (whether it is air or water). The air storage should be determined by finding the sizes and number of air receivers from the plant personnel or from the tank, and the diameter and length of the header pipes should be measured. The number of compression stages and lubrication-type information should be given by plant personnel or can be found on the Compressed Air and Gas Institute (CAGI) data sheets [12]. CAGI data sheets may be produced by the manufacturer that develops the air compressors and can be found on the manufacturer's website using the model number of the compressor, which is found on the nameplate. For centrifugal compressors, a performance map that is created by the manufacturer is needed for the analysis of centrifugal compressors [12].

The last and most important information that is needed for the analysis of compressed air systems is the input power or airflow data. This current flow data can be provided by the plant personnel if the plant has a recording system. If the plant does not have such a data-logging system in place, current transducers with data loggers connected to them can be attached to the input power lines in the power cabinet on the compressors, or at the power disconnect, to measure the current draw of each compressor in the system [10]. Plant safety rules have to be taken into account before opening the power cabinet or disconnect box for each compressor. At least a week's worth of data needs to be measured (though more is better) to be able to model the compressed air system accurately.

Consistent with the American Society of Mechanical Engineer (ASME) protocol [13], the analysis of the assessment data begins by using the collected data and information to create a baseline profile of the compressed air system. The baseline profile should include power and energy profiles, air-demand profiles, supply efficiency, identify the different operating period types, and should include annual air demand and energy consumption. The annual energy consumption can be estimated by using the number of each type of compressor, the motor horsepower, the motor efficiency, the load factor, and operating hours [12]. After that, these numbers can be multiplied together to get the annual energy consumption of the compressed air systems. The system volume (effective volume) should also be calculated, along with a pressure profile of the system. The high-pressure demands on the system should be validated to be sure that the high pressure is required. Along with the pressure profile, an air-demand profile should also be created [13]. The critical air demands and the wasted compressed air should be analyzed. Critical air demands have to be met to assure that high product quality is being repeated, and the compressed air waste leaks, inappropriate uses, and artificial demand that can decrease the efficiency of the system. The air treatment equipment should be examined for optimization. The team should establish a target pressure for the system to increase the efficiency of the compressed air system [11]. Balancing the supply and demand is another recommendation that should be investigated for an increase in energy efficiency for the compressed air system. Assessing the maintenance opportunities and evaluating the heat-recovery opportunities are included in the final analyses.

3.2. Recommended assessment opportunities

The first and most common recommendation is lowering the overall system pressure in the compressed air system [10]. Discussion with facility personnel will help determine the highest needed pressure for the process equipment. By looking at the gauges on the compressor, the display screen on the compressor, and the flow control system that is in place, the operating pressure can be determined and can be compared with the needed pressure of the process equipment. This recommendation can be achieved by progressively reducing the discharge pressure at the compressor using the compressor controls, or by using a flow controller to equal out the system pressure at the required set point for the plant [11].

The next most common recommendation is reducing the amount of leaks in the compressed air system. An air-leak survey should be executed using an ultrasonic air leak detector, and the decibels of each leak should be recorded. Also, if power or airflow data can be recorded for times without any production in the facility, then a leak load can be found from this power or airflow data during this period. This assessment recommendation can be completed by implementing a maintenance program to check for compressed air leaks on a regular basis to keep the percentage of leaks down. Completely reducing the leaks in a compressed air system to zero is nearly impossible, especially for large systems, but with a well-implemented leak program, leaks in the system can be reduced to 10% of the average airflow of the system [11].

Another indication of a possible AR is the existence of inappropriate usages of compressed air, for example, tank sparging, part cleaning, and drying, which should be reduced or eliminated [13]. Discussion with facility personnel can help to determine some inappropriate uses that can be performed more efficiently with another energy source. For example, using compressed air for tank sparging (i.e., to mix up liquids) is very inefficient compared to using a pump or a stirrer. Using compressed air for part cleaning or drying is similarly inefficient compared to using some type of a blower system.

Using an automatic sequencer is another potential energy-saving AR. The need for this is indicated by having multiple compressors on a system without any automatic control, and only having manual control by facility personnel. Automatic operation controls will rotate compressors in and out of the system as needed and will alternate the backup compressors into the system [11].

Recovering waste heat from the compressor is another potential energy-saving AR. Indications of the need for this AR are having air-cooled compressors venting air to atmosphere, as well as a need for heat recovery in some process or space conditioning in the plant [14]. Recovering waste heat can improve the efficiency of a system that requires heat, but this AR does not increase the efficiency or reduce the energy consumption of a compressed air system. If the compressors are air-cooled and are located inside a conditioned space, then venting the heat out of the conditioned space during the summer months and into the conditioned space during the winter months will assist the HVAC system [14].

Using an optimum-sized compressor is another potential energy-saving AR. A compressor operating at the low end of its operational range is an indicator for this alternative [13]. An oversized compressor can be determined by measuring the electrical current flow and comparing the value to the motor nameplate data. The power should be plotted on the motor curve to check to see if the motor is operating at a high efficiency point. Using an oversized compressor motor at partial load will not be operating near its highest efficiency point, and therefore will be using more power than needed to produce the compressed air [12]. To calculate the optimal-size compressor, the required pressure and flow rate are needed to compare other compressor characteristics [11]. The energy savings will result from using less power to produce the required pressure and flow rate.

Another possible energy-saving AR regards using a dedicated air compressor. This may be indicated by having an end use at a considerable distance from the compressors, or having an end use that requires a higher pressure than the rest of the end users. Utilizing a dedicated compressor for end users with high-pressure requirements will reduce the cost to produce compressed air by making the lower-system pressure AR, discussed previously, viable [11]. The required airflow and pressure of the process equipment, which needs the dedicated compressor, are required to size the compressor properly.

Installing a variable frequency drive (VFD) compressor is another potential energy-saving AR. Variable frequency drives (VFDs, alternatively referred to as variable speed drives or VSDs) change the speed of the motor to increase or decrease the amount of power consumed, which is proportional to the output flow capacity. These drives can control the output flow capacity of a compressor from 15 to 100% of full flow [15]. Anything below 15% of full flow can result in the compressor being unloaded or shut off. The power factor of the motor while using a VFD is normally better than other conventional controls, and VFDs can yield a constant pressure band. Typically, compressors that are originally designed for VFDs have a higher benefit while using VFDs than compressors that have been retrofitted with VFDs [15]. An indicator for the need for this AR is having a variable load from the end users in a compressed air system, or a compressed air system that needs a trim compressor [10]. Using a VFD compressor will use less power more efficiently if the compressed air system has a variable load when the load is less than the full load of the compressor. Variable speed compressor curves are needed for the compressor to help determine the energy savings of a VFD [13].

Reducing the run time of the air compressor may also be considered, particularly if the compressor is being operated during non-production hours [10]. Setting the compressor controls to turn off when the compressed air system is not needed will save energy. The amount of time that the compressor can be shut off is the primary data needed to determine the resulting energy savings.

http://dx.doi.org/10.5772/64902

If the compressor intakes are in locations where the ambient air temperature is high, then another possible AR is installing compressor intakes in the coolest location possible [12]. If the compressor is drawing air from an air-conditioned plant, then savings can be found by using outside air to reduce the load on the HVAC system. However, the energy savings or efficiency increase for compressors using cooler intake air is not easily calculated and is being researched by IAC personnel at this time [10].

4. Boilers and steam systems

Steam, along with electricity and direct-fired heat, is one of the three principle forms of energy that is used in industrial processes. Steam can range from 28 to 76% of the total onsite energy depending on the type of industry [16]. Boiler systems are divided into four different subsystem categories that include the generation, distribution, end uses, and recovery of steam. The steam is created with the generation components, which include boilers, pumps, and economizers. Once the steam leaves the boiler, it flows through the distribution system, which contains pipes, valves, and backpressure turbines. An efficient distribution system provides the appropriate amount of steam at the right temperatures and pressures to each end use. Steam can be used for numerous different processes and applications. Some end uses of steam are for process heating, mechanical drive, chemical reactions, and separation of hydrocarbon components. End-use components include heat exchangers, turbines, strippers, chemical reaction vessels, and fractionating towers. Finally, after the steam is used by the end uses, the condensate return system captures the condensate and sends it back to the boiler. The condensate is sent to a collection tank or to a deaerator tank to mix with the makeup water before the feedwater is pumped into the boiler. A deaerator tank is a vessel that is used to reduce the oxygen content in the boiler's feedwater. Deaerator tanks pressurize the feedwater, and the temperature is increased to the point of saturation, along with removing oxygen and other non-condensable gases [16].

Heat recovery, from the flue gases of natural gas or fuel oil boilers, can increase the efficiency of boilers by preheating the feedwater before it goes into the boiler. This heat can be recovered by heat exchangers in the exhaust stack of the boiler, and are referred to as economizers. In general, economizers usually can reduce fuel requirements by 5–10% [14].

Stack economizers are gas-to-liquid heat exchangers that are installed into the exhaust stack of the boiler, to recover sensible heat from the flue gases of natural gas or fuel oil boilers. Stack economizers can only recover sensible heat from the flue gases and can reduce the flue gas temperature down only to about 250 F (or 121°C). If the flue gases are reduced to a temperature below this value, then condensation can develop in the exhaust and this can decrease the life of the economizer [17]. Stack economizers can contain bare carbon-steel tubes or finned tubes depending on heat-recovery targets and the composition of the flue gases [14]. Condensing economizers can be used on large or small boilers and can be an attractive energy efficiency measure when stack economizers are not. Condensing economizers are heat exchangers that can recover sensible and latent heat from the flue gases of a natural gas boiler. More heat can be recovered using a condensing economizer than using a stack economizer [16].

4.1. Recommended assessment procedures

After the kick-off meeting and facility walk-through, the target equipment and components should be evaluated by measurement equipment, and identification and collection of essential data for the systems should recorded [10]. These essential data include temperature measurements of boiler makeup water, feedwater, shell, ambient air, stack gases, steam headers, and the distribution piping. The required data for pressure measurements include steam headers and branch lines, condensate return tank, deaerator, and points of usage before pressure reduction valves [3]. The flow measurement data that are required are the boiler fuel input rate, steam output rate, makeup water, blowdown, and the consumption of the end uses. Finally the last set of data that are needed is the chemical measurements (conductivity), which include the chemical concentrations (dissolved solids, chloride, and silica) for the makeup water, internal boiler water, condensate, and feedwater [16]. Once all of the data are measured and recorded, a system baseline can be established.

The number of different headers and the steam pressure at each header should be recorded in the notes, along with the process of each pressure reduction, for example, whether it is throttled or runs through a turbine. The fuel consumption is critical information that can be identified on a meter or from plant personnel. Similarly, the feedwater usage, obtained from a meter or from plant personnel, is also valuable information that is required. Other needed data include the percentage of condensate return, makeup water usage, the blowdown rate, the deaerator tank vent percentage and pressure, the feedwater usage, the load factor of the pump, the firing rate of the boiler, and the exhaust temperature and oxygen content in the exhaust stack [16]. The total number of steam traps on each header should be known, and the number of failed steam traps on each header, along with the number and size of leaks in the distribution system and the heat losses from the distribution system, should be identified [17].

If there is an economizer present in the system, it should be noted, along with the application that it is used for, and the exhaust temperatures before and after the economizer should be measured. The isentropic efficiencies of the steam turbines should be estimated, if a steam turbine is present in the system. The isentropic efficiency of a steam turbine can be calculated by dividing the actual turbine work by the isentropic turbine work [18]. The inlet and outlet pressures and temperatures of the steam should be measured, and then using the steam tables the enthalpy at each state can be found. Once the enthalpy at each state is found, the actual and isentropic turbine work can be found [19]. If the pressure and temperature cannot be measured, then the isentropic efficiency can be assumed to be between 70 and 90%. Where the larger turbines are closer to 90% and the smaller turbines are closer to 70%. Occasionally well-designed, large turbines can have an isentropic efficiency above 90%, and small turbines can be less than 70% if the turbine is not designed well [19].

During analysis of the assessment data, a final steam system baseline should be created using the collected data from the assessment visit. The annual energy consumption can also be estimated by using the number of boilers in each system, nameplate data (the rated input), the load factor, and operating hours of the boilers [10]. After that, these numbers can be multiplied together to get the annual energy consumption of the boilers. Once the baseline is created, then the energy-saving recommendations can be developed.

http://dx.doi.org/10.5772/64902

4.2. Recommended assessment opportunities

Recovering waste heat in the stack of the boiler presents one potential energy-saving AR. The exhaust gas analyzer should be used to take a measurement from the flue gases in the exhaust to determine this temperature. The higher the exhaust temperature the more heat can be recovered [14]. This recovered heat can increase the efficiency of a boiler system, process heat system, or HVAC systems. The heat is most commonly recovered by a stack or a condensing economizer and is used for preheating boiler feedwater [14].

Reducing the amount of leaks in the steam system via implementing a maintenance program is another potential energy-saving AR. Steam leaks are quite noticeable while observing the steam system. With normal circumstances, leaks can be minimized if a well-implemented maintenance program is used [1]. The sizes and number of leaks are the main data that are needed to determine the energy savings for decreasing the amount of leaks in a steam system [3].

When the oxygen content in the exhaust gases is higher than 3%, another potential energy-saving AR is reducing this oxygen content [16]. The exhaust gas analyzer should be used to take a measurement from the flue gases in the exhaust to determine the oxygen content in the flue gas. Reducing the oxygen content in the exhaust gases will increase the efficiency of the boiler. Frequently, installing an electronic oxygen trim control system on the boiler will reduce the oxygen content [18].

Exposed piping, valves, and fittings in a steam system's piping network indicate the need for addition insulation to these exposed surfaces. Taking thermal images will show the hot spots in the steam distribution system and will indicate the temperature of the hot spot [10]. Insulating the piping, valves, and fittings will decrease the heat loss of the steam system, therefore increasing the efficiency of the steam system. Similarly, insulating hot spots on the boiler surface will decrease the heat loss of the boiler, therefore increasing the boiler's efficiency.

Reducing the boiler pressure is another possible energy-saving AR. Steam produced at higher pressure and throttled for all end uses provides evidence for this potential energy-saving AR [12]. The facility personnel can provide the boiler pressure, and the highest pressure that is used by the processes. Reducing the boiler pressure will decrease the amount of work needed to produce the steam.

When steam is produced at higher pressure and throttled for all end uses, an alternative energy-saving AR involves installing a steam turbine into the steam system [18]. A steam turbine will produce electricity while replacing the throttling valve and decreasing the steam pressure in the system for the processes. The main required data are the current boiler pressure, steam pressure needed by the process, and the process of pressure change between the boiler and the next-level header.

Replacement of failed steam traps, as well as implementing an associated maintenance program, presents the final type of energy-saving AR. Thermal images will indicate which traps are working and which have failed [10]. Failed steam traps do not let the steam bypass back to the steam system. Replacing failed steam traps will increase the amount of condensate returned to the boiler.

5. Process-heating systems

Process heating is essential in the manufacture of most industrial and consumer products. Process-heating equipment is either fuel-fired, electric-based, or steam equipment. In the United States, fuel-fired process-heating equipment consumes approximately 17% of the total industrial energy consumption [20]. Process-heating equipment has many different names, which include furnaces, kilns, heaters, ovens, lehrs, incinerators, melters, and dryers, but all basically operate under the same rules as just heating a load to complete a task [20]. Heat-treating furnaces are used to create mechanical properties of metals, which includes strength, hardness, and flexibility. Heat-treating furnaces are used in the metal production industry mostly, but also in industries that anneal and temper ceramics and glass [20]. Drying ovens are used for water removal through direct or indirect heating. These ovens are used in industries that need dry raw materials or finished product that contains water. Drying ovens are common in glass, clay, food processing, textile, and chemical industries [20].

5.1. Recommended assessment procedures

Operational and maintenance requirements need to be identified by the assessment team members. Practical requirements of each process-heating system include the energy usage and emissions, production output, and the quality of the products [3]. If there are meters that measure the flow rate of the fuel and flue gas oxygen content, then these data should be recorded. Also, if any electrical power meters are installed on any equipment, this information should be recorded. Some facilities may have data-recording systems that can produce data that are needed for the energy assessment. The control systems and control strategies should be determined for each process-heating system. These can include the control of the heat input, temperature, and the air-to-fuel ratio.

The furnace type and fuel type of the furnace will need to be recorded, along with the number of burners for each furnace in each system. The nameplate data of the furnaces are located on the furnace or burners and should be recorded in the assessment notes, and a picture should be taken of the nameplate if available, consistent with the ASME energy assessment standard [21]. Plant personnel should be able to explain how the schedule of the process-heating system is set up; the operating hours along with the non-operating hours are critical to assessing the energy usage.

The nature of the material comprising the load or charge is very important for a process-heating system energy assessment. Along with the type of material, the feed rate of the load, initial charge temperature, and discharge temperature are also needed [21]. Next, an exhaust gas sample needs to be measured using a flue gas analyzer. This measurement instrument has a probe that needs to be inserted into the exhaust stack as close to the top of the furnace as possible, and should make a complete seal with the port on the exhaust to avoid dilution air in the exhaust [10]. The gas analyzer measures the exhaust gas temperature, ambient air temperature, and oxygen content of the exhaust gas, and then calculates the excess air, combustion efficiency, and carbon dioxide content of the exhaust gas.

After the load and exhaust gas information is found, the losses in the furnaces need to be determined and recorded [21]. To find the fixture losses, the fixture material, weight, initial temperature, and final temperature should be measured and recorded. The information that needs to be measured to find the losses in the walls includes surface area of the walls, average surface temperature, and the ambient temperature. For furnaces with special atmospheres inside the furnace for certain processes, the type of gas needs to be determined. Along with the type of gas, the initial temperature, final temperature, and flow rate of the special atmosphere gases need to be determined. Opening losses are one of the most common losses with a high magnitude [18]. To determine the losses through openings in the furnace, the type, shape, and size of the openings are needed along with the furnace wall thickness, inside temperature, ambient temperature, and percent of time open. For furnaces that use water for cooling, some losses will arise from this. To find the water-cooling losses, the water flow rate, inlet temperature, and outlet temperature are needed [3]. Finally, the heat storage of each furnace should be assessed and recorded. The furnace shape (rectangular or cylindrical), furnace size (height or diameter), furnace temperature, ambient temperature, and starting wall temperature are all required to determine the heat storage of the furnaces [3].

When analyzing the assessment data, each process-heating system requires an energy balance. This baseline should be portrayed in units of energy per production unit or energy per unit of time for each process-heating system [21]. The annual energy consumption can be estimated by using the number of furnaces in each process-heating system, the rated input for the burner, the load factor, and operating hours of the furnaces. After that, these numbers can be multiplied together to get the annual energy consumption of the furnaces [20]. Next, the assessment recommendations should be identified. These recommendations can include maintenance improvements, operation enhancements, control strategy upgrades, process improvement changes, and equipment replacements.

5.2. Recommended assessment opportunities

Consistent with the discussion of boilers in Section 4 of this chapter, heat recovery from the flue gases of fuel-fired process-heating equipment can increase the efficiency of furnaces by preheating the combustion air before it goes into the burner. In this case, heat can be recovered by heat exchangers in the exhaust stack of the boiler, referred to as air preheaters. In general, furnace efficiencies can be increased by approximately 20–30% for fuel-fired furnaces [14]. There are two types of air preheaters: recuperators and regenerators. A recuperator is fixed heat exchanger in the exhaust stack of a furnace. This air-to-flue gas heat exchanger is used to preheat combustion air using the flue gases. A regenerator is a container that is insulated and filled with metal or ceramic shapes that absorb thermal energy. Regenerators can store a moderately large amount of this thermal energy and then release that energy subsequently to preheat the combustion air of a furnace [14]. The exhaust gas analyzer should be used to take a measurement from the flue gases in the exhaust to determine the flue gas temperature. The higher the exhaust temperature the more heat can be recovered. This recovered heat can increase the efficiency of a process heat system, boiler system, or HVAC systems. The exhaust gas sample, preferably the exhaust temperature, is used to determine the type of heat-recovery method to be used, and then the energy savings will be calculated by finding the amount of heat that can be recovered from the exhaust temperature [14]. Once the amount of recovered heat is calculated, then the increase in combustion efficiency can be found for purchasing a heat exchanger and installing it in the furnace's exhaust stack.

Also consistent with the discussion of boilers, another potential energy-saving AR is reducing the oxygen content in stack gases. The exhaust gas analyzer should be used to take a measurement from the flue gases in the exhaust to determine the oxygen content in the flue gas. An indication of this AR is having oxygen content in the exhaust gases higher than 3% [18]. Reducing the oxygen content in the exhaust gases will increase the efficiency of the furnace.

Insulating the hot spots on the furnace walls is another potential energy-saving AR when such hot spots are observed via thermal imagery on the furnace walls [10]. This may be due to bare or inadequately insulated surfaces. Insulating these hot spots will decrease the heat loss of the furnace, therefore increasing the furnace's efficiency. Reducing the amount/size of openings to the atmosphere, in the furnaces, will similarly provide another possible energy-saving AR. These openings may be observed while examining the process-heating equipment. Reducing the area of these openings in a furnace will also reduce heat loss, and therefore increase the furnace's efficiency. The implementation of these last two ARs requires the purchasing and installing of insulation on the furnace walls.

6. HVAC systems

In 2003, heating, air conditioning and ventilation (HVAC) systems consumed about 30% of the energy consumption for commercial buildings. Space cooling represented about 44 of the 30% consumed by HVAC systems, while space heating and ventilation represented about 16 and 40%, respectively [22]. HVAC systems typically have one or more of four basic types of units to produce conditioned air, along with a duct system for the air distribution to the facility or a particular area in a facility. These four types of units are packaged units, air-handling units, split system air-conditioning units with gas furnaces, and split system heat pump units with auxiliary heat [23]. All HVAC units contain fans, filters, and coils. HVAC systems are used for cooling and heating of facilities, and keeping the humidity of an area or facility to a required level (since some processes have certain humidity or temperature requirements within a facility).

6.1. Recommended assessment procedures

A preliminary energy-use analysis (PEA) should be completed before any level of audits [24]. The PEA provides the essential background information for energy assessments of any level. It includes the following steps: defining the floor area of the facilities' conditioned space and recording the floor area. The next step is to collect at least a year's worth of utility bill data, and these data should be summarized to look at opportunities to change the rate schedule, if applicable. For the PEA, the utility use, peak demand, and costs should be analyzed, along with developing the energy cost index (ECI) of the building, which should be conveyed in dollars per floor area per year [24]. The energy utilization index (EUI) should be developed during the PEA, which should be expressed in energy use per floor area per year. The energy performance summary should be completed to develop the ECI and EUI for each energy (fuel) type and demand type. Once the ECI and EUI are developed, then these indices should be compared to similar buildings that contain comparable characteristics [24]. After the comparison is made, then the new energy, demand, and cost goals should be established, and then using the new values, calculate the energy and cost savings for each fuel type.

According to the American Society of Heating, Refrigerating and Air-Conditioning Engineers (ASHRAE) [24], a level 1-walk-through analysis of HVAC systems consists of first performing a brief walk-through survey of the facility to become familiar with the building's construction, operation, equipment, and maintenance. The next step is meeting with facility personnel to learn of special problems or scheduled improvements and any maintenance issues that affect the overall efficiency of the HVAC systems. Subsequently, a space function analysis should be completed to determine whether or not the HVAC system efficiency has decreased due to different functions in the building [24]. If the current functions in the facility are different from the old functions, then the HVAC system may not be designed correctly. Finally, the energy reduction opportunities should be identified. These opportunities should be split into two categories: the low-cost or no-cost opportunities, and the capital investment opportunities. An initial rough estimate of energy and cost savings should be made for the opportunities.

The HVAC unit type and manufacturer should be recorded in the assessment notes along with the nameplate data of each unit, and a picture should be taken of the nameplate, if possible. The nameplate should be on the side of the outside panels of the unit. The thermostat set points should be recorded, facility personnel should be able to tell what these set points are, but it is good practice to double check by looking at the thermostats [24]. Identification of the types of thermostats that are used is important, that is, whether it is a basic thermostat or a programmable thermostat needs to be recorded. The plant could have a building management system that can control the set points automatically [23]. Plant personnel should be able to explain how the schedule of the HVAC systems is established; however, most of the time the operation only has to do with the set points and outdoor temperature. The operating hours for the HVAC systems can be found by using historical weather data for the region that the facility is located.

The fan sizes should be recorded if not on the nameplate data. The most important information that is needed for the analysis of the HVAC systems is the input power data for each HVAC unit. If the plant does not have a data-logging system in place, current transducers with data loggers connected to them can be attached at the power disconnect, to measure the current draw of each unit, as discussed in previous sections of this chapter. Also as noted previously, facility safety rules must be considered before opening the disconnect box for each unit. At least a few days of data need to be measured, a week's worth of data or more would be better, to be able to model the HVAC system accurately [23]. This power data will be used to calculate the energy usage of the HVAC systems.

The airflow rate data should be measured in the duct system, in different places, to determine if any major losses are affecting the airflow. The flow rate can be measured by a pitot tube in the duct, or a vane anemometer at an outlet of a duct. According to the Air Movement and Control Association International [25], measuring the flow rate using a pitot tube requires a specific technique, and is similar but not exactly the same for circular ducts and square ducts. These measuring points should be located at the center of equal area squares in the duct. The cross section of the duct should be divided into equal area squares, and the amount of squares should be equal to the amount of measuring points needed [25]. For good results, the pitot tube device should be placed in the duct at least 10 duct diameters away from the last fitting (or anything that creates a loss in the duct system) [25]. The facility duct system layout and duct diameters should be recorded, along with any places that have a potential loss in the duct network.

The uses of the system should be recorded along with the temperature and humidity requirements of each zone in the facility [24]. The annual weather data for the region should be researched and can be found online from the national weather service and other websites. The building construction materials used need to be described, and this information can be provided by facility personnel or building specification sheets [24]. The team members can measure and record the type of walls in the building, along with measuring the building dimensions. The sources of heat in the building should be recorded in the assessment notes. These sources of heat can include processes, machinery, furnaces, boilers, lighting, facility personnel, HVAC systems, air compressors, building envelope, infiltration, and plug loads.

Once all of these data are gathered, the analysis can calculate the annual energy consumption of the HVAC systems by using the number of each type of unit, the tonnage of the unit, the coefficient of performance (COP), the usage factor, and operating hours [23]. Subsequently, these values can be multiplied together to obtain the annual energy consumption of the HVAC systems.

6.2. Recommended assessment opportunities

Implementing a HVAC unit maintenance program is a potential energy-saving AR, if one does not already exist. Many factors can reduce the efficiency of the unit, and these factors include a reduction in evaporator airflow, refrigerant line restrictions, and refrigerant undercharging or overcharging [25]. Implementing a HVAC unit maintenance program will keep the units in top condition and will result in a lower efficiency loss over the lifetime of the unit. A maintenance program will also increase the lifetime of each unit.

Utilizing airside economizers provides another possible energy-saving AR. Airside economizers are able to take cool dry air from the outside atmosphere and use it to cool the facility. These economizers can be operated with either a temperature control or an enthalpy control to determine when the outside air is suitable for cooling [23]. The use of airside economizers will result in less power required for cooling because the compressor will not have to operate as often.

Another potential AR for consideration is the use of programmable thermostats and the subsequent adjustment of set points to an optimal temperature. Observation of the thermostat set points being lower in the summer months and higher in the winter months than needed, as well as having non-programmable thermostats, provides evidence supporting selection of this AR [3]. Adjusting the thermostat set points to an optimal setting will reduce the operational hours of the HVAC system, therefore reducing the energy consumption. Using programmable thermostats will allow a schedule to be established for the set points to be dialed back during nonworking hours, which will reduce the run time of the HVAC units while the facility is vacant [24].

Applying a new roof coating to the facility is another possible energy-saving AR. A new roof coating can reduce the run time of the HVAC system by increasing or reducing the heat gain of the building envelope [23]. For colder climates, where heat is needed the most, a black roof coating can be beneficial. However, for hot climates, a white roof can be more beneficial to reduce the heat gain. The implementation of this AR requires replacing the roof coating, by just simply either painting over existing coating or removing old material and replacing with new roofing material.

Adding insulation to the building is a potential energy-saving AR. Thermal images of the inside and outside of the building walls will help determine whether this AR is viable or not. Adding insulation to the building will reduce the heat gain into the building during the summer and reduce the heat losses in the winter [23]. Therefore, the HVAC system will have to operate less and decreases the overall energy consumption.

A building that needs dry/dehumidified air, rather than cool or warm air, should consider installing a desiccant dehumidification system. Discussion with the facility personnel about the temperature and humidity requirements for the areas in the building can support the practicality of this AR. Installing a desiccant dehumidification system can take air and reduce the humidity to a point that is suitable for the processes inside a building [24]. Desiccant units can be used in place of HVAC units to dehumidify the air while using less energy. Since these desiccant dehumidification systems do not heat or cool the air, and if special temperature requirements for the manufacturing process need to be met, then this AR may not be viable [23].

7. Pumping systems

Industrial motor systems are the single largest electrical end-use category in the United States, and pumps account for about 27% of the industrial motor energy consumption [26]. A pumping system contains one or more pumps with motors and a piping network that includes valves and fittings. These pumping systems are categorized as either closed-loop or open-loop systems. Closed-loop systems recirculate the water that is contained in the piping network, and open-loop systems contain a sump where the liquid is pulled from and the water is either discharged back into the sump or to the needed process. Pumping systems are used for facility HVAC processes and facility production processes [26]. In HVAC systems, heat exchangers use water to transfer heat to air for space conditioning, and for heat exchangers the flow rate from the pump is the critical performance characteristic. Process equipment use pumps to provide hydraulic power to machines, where pressure is the critical performance characteristic that is needed [26].

7.1. Recommended assessment procedures

An ASME level 1 assessment should include gathering information for each pumping system that is in the scope of the assessment [27]. The prescreening should include listing of the pumping systems, pump type, motor nameplate data, annual operating hours, applications of the pumps, and the control methods. During the prescreening, the systems that should be evaluated more closely should be determined, and any systems that can affect other systems should be noted to present the constraints on the systems [27]. The prescreening process should sort the systems by size, energy costs, and operational hours. Fixed speed centrifugal pumps and systems with throttling, recirculation, or by-pass controls should be a main focus of the assessment [3].

Recording the nameplate data of the pump and motor is a protocol of the ASME energy assessment standard [27]. The pump type and brand of the pump should be recorded also, which may be on the pump housing or on the nameplate. Plant personnel should be able to indicate the operating hours along with the non-operating hours, which are critical to assessing the energy usage [3].

Primary information that is needed for the analysis of the pumping systems regards the input power data. The procedure for obtaining these data follows the pattern from previous sections of this chapter. The next most important data to obtain while on an assessment are the flow rate data for the fluid in the system. Flow rate data can be given by the plant personnel if the plant has a recording system that properly records data or be taken from a flow rate-measuring device that is already in place on the piping network. If no flow rate-measuring device is in place, then an ultrasonic flow rate meter can be used to measure the flow rate [10]. This device can measure the flow rate in pipes of various diameters. For good results, the device should be placed on the pipe at least 10 duct diameters downstream from the last fitting, valve, or anything that creates a loss in the piping system and five duct diameters upstream [27]. Recording flow data this way is consistent with the ASME standard for measuring flow data. Pressure data should also be recorded at various points in the system, especially at places where large pressure drops could occur. Measuring pressure can be hard to accomplish if there are not any pressure measuring devices in the system already. If that is the case, then asking plant personnel may be the best way to get the pressure in the system. Data for calculating the system head are also needed, including the elevation difference in the piping network, pipe diameter, and the losses in the piping network [26]. The elevation difference and pipe diameter should be measured, and the losses should be counted and recorded along with the type of loss (valve, fitting, etc.). Pressure and head data are both an ASME standard [27]. The end uses of the system should be recorded along with the amount of fluid needed for each use.

The last piece of information needed to analyze a pumping system is the pump curves for each pump. These curves can be found on the manufacturer's website using the model number of the pump, or by contacting the manufacturer with the model number which is found on the nameplate. Pump curves include a head versus flow curve at particular impeller diameters, efficiency curves, power curves, and net positive suction head (NPSH) curves [26].

From the collected data, the ideal amount of energy required for the system to achieve the essential functions should be calculated by using the number of each type of pump, the motor horsepower, the motor efficiency, the load factor, and operating hours. Once the optimal amount of energy is calculated, then recommendations to increase system efficiency can be calculated. The system curve should be calculated for each pumping system in the facility. Only two points are needed to generate a system curve: the head at zero flow or static head point, and one operating point [27]. The system curve is required to completely understand the pumping system. As each recommendation is evaluated, a new system curve should be developed before deciding on the next recommendation. The energy-saving recommendations include reducing the system head, reducing the flow rate, confirming that pumping system components are operating close to the best efficiency point, and changing the operating hours of the pumping system. An optimal pumping system energy profile should be determined using the best recommendations, while the system requirements are reached [27].

7.2. Recommended assessment opportunities

Downsizing the existing pumps in a pumping system is a potential energy-saving AR. An indication of the possible AR is having a pump lightly loaded or flow throttled to desired rate. An oversized pump in a pumping system can be found by measuring the power or motor speed and plotting on the pump curve to check to see if the pump is operating at a point with high efficiency [27]. Using an oversized pump will use more power than needed to perform the job of the system.

Installing a VFD pump in a pumping system is a possible energy-saving AR for consideration. In the compressed air section of this chapter, VFDs are defined. Pumps use VFDs to increase efficiency and save energy by matching the flow rate to the requirements of the process. VFDs can be considered to be 95% efficient for pumps from 25 to 100% of full load [28]. VFDs on pumps offer minimum energy consumption over the entire speed range of the drive [29]. Evidence of the need for this potential AR is having a variable system flow with control by throttling or by-pass valve. If a flow control system (throttle valve or by-pass) is in place, and the controls are restricting the flow by more than 40% (60% closed), then VFD controls will result in energy savings [28]. Using a VFD will use less power more efficiently if the pumping system has a variable load.

Installing an automatic control in a pumping system is another potential energy-saving AR. Automatic operation controls will rotate pumps in and out of operation as needed by the system, and will alternate the backup pumps into the system [26]. The energy savings will vary from system to system depending on the quantity and size of the pumps in the system. The energy savings are based on the resulting reduction in power and operating hours in a pumping system. The implementation of this AR requires the installation of an automatic control system into the existing system, which will require sensors installed into the system including flow rate meters, temperature sensors, and pressure gauges [26].

Shutting off unneeded pumps in a pumping system is also a possible energy-saving AR. An indication of need for this AR is observing several pumps in parallel operating below their best efficiency point. An unneeded pump in a pumping system can be found by measuring the electrical current or motor speed, and plotting on the pump curves [26]. Unneeded pumps can be observed running at a very low efficiency, or even at dead head (which means the pump is operating at zero flow and only recirculating water in the pump housing) [27]. This can result in the pump overheating and subsequent damage. Shutting off unneeded pumps will reduce the power consumption of the pumping system. The energy savings will also vary from system to system depending on the quantity and size of the pumps.

8. Fan systems

Industrial motor systems are the single largest electrical end-use category in the United States, and fans use about 78.7 million kilowatt-hours of energy per year [30]. This 78.7 million kWh accounts for about 15% of the industrial motor energy consumption. Fans are used in nearly every manufacturing facility and are used for various HVAC purposes and facility process purposes [30].

Ventilation fans are one of the most common uses of fans in industrial facilities. They are used for heat control during the summer; the fan creates a draft through the plant to remove the heat from workers and machines. Local exhaust systems are designed to remove hazardous fumes or contaminates from chemical or mechanical process inside of a facility. They are generally located above or beside the process with a hood, and the fumes or contaminates are ducted out of the process area. These exhaust systems can include filters and dust collectors. Baghouses use fans and cloth bags to filter out dust particles by pulling air through the cloth bags to filter out the unwanted particles [31]. Another industrial application of fans is mechanical draft systems that move air through boilers and furnaces. Mechanical draft fans supply combustion air, deliver fuel to the burner, circulate gases for better heat transfer, and remove combustion products [31]. Air-blast drying and air-blast cleaning are two more applications of fans in an industrial facility. Parts that have been washed need to be dried and this can be done by air-blast drying. Parts that need particles cleaned off of them can be put under an air-blast cleaning fan to remove the excess or foreign particles [31]. Personnel cooling is the frequent use of fans other than ventilation fans in facilities. These fans range from small axial fans that are spread throughout a facility at workstations, or high volume low speed (HVLS) that are large ceiling fans that can cool large areas inside a facility.

Every HVAC system has to contain one or more fans to distribute the airflow throughout the zone that the system conditions. Packaged units and split system units have axial fans on the condenser side to pull air through the condensers to cool off the refrigerant and have centrifugal fans for the supply air fan that forces the air into the duct system [23]. Fan performance can be degraded due to pressure drops across filters and heat exchangers, air density can also alter the performance of a fan system [30].

http://dx.doi.org/10.5772/64902

8.1. Recommended assessment procedures

The nameplate data of the fan and motor are located on the respective housings, and should be recorded in the assessment notes, and a picture should be taken of the nameplate if available. The fan type, brand, and operating hours of the fan should be recorded also. The annual energy consumption can be estimated by using the number of each type of fan, the motor horsepower, the motor efficiency, the load factor, and operating hours [32]. Each duct system layout and duct diameters should be recorded, along with any places that have a potential loss in the duct network.

Similarly, the flow rate data for the gas in the system need to be recorded. A vane anemometer or a pitot tube can be used to measure this flow rate. The vane anemometer can measure the exit velocity in ducts of various diameters, and the pitot tube can also measure the flow rate anywhere in the duct system [25]. The procedure for measuring the flow rate and pressure data, using a pitot tube, is discussed in the HVAC section of this chapter.

The last piece of information needed to analyze a fan system is the fan curve for each fan. Similar to the pump curves discussed in the previous section, fan curves can be found on the manufacturer's website using the model number of the fan or by contacting the manufacturer with the model number which is found on the nameplate. Fan curves include a static pressure versus flow curve, power curves, and fan curves for different speeds can also be found [30].

8.2. Recommended assessment opportunities

Observing an excessive flow control to reduce volume indicates consideration of downsizing the existing fans as a potential energy-saving AR. An oversized fan in a fan system can be found by measuring the electrical current or motor speed and comparing them to the fan curves [30]. The power should be plotted on the fan curve to check to see if the fan is operating at a point with high efficiency. Using an oversized fan will use more power than needed to perform the job of the system.

Installing a VFD fan in a fan system is a possible energy-saving AR. Fans use VFDs to increase efficiency and save energy by matching the airflow to the process requirements and by being able to operate over a wide range of operating conditions [30]. VFDs are used for fan system controls when the load on the system is frequently changing or a continuously variable load is present [32]. VFDs can modulate the speed of a fan from full-design speed (100%) down to 20% of design speed [15]. Since VFDs can keep fan efficiencies high across many variations in the system, the operating costs can be reduced [30].

Replacing standard v-belts with cogged v-belts, on existing fan drives, is a possible energy-saving AR. Using a cogged belt can increase the drive efficiency of a fan by 2–3% [33].

Replacing the personal fans in the facility with high-volume low-speed (HVLS) fans is also a potential energy-saving AR. Using small personal cooling fans can add up if every employee or every workstation has one or more fans; HVLS fans use a small efficient motor to generate a high volume of air movement. Depending on the amount of area, the placement of the lights, and the placement of other objects close to the ceiling, a significant problem can be seen with the implementation of this AR. Lights or other objects that hang down far from the ceiling have the potential to be an obstruction that can make installing HVLS fans difficult [10].

9. Conclusions

Industrial energy assessments form a fundamental segment of energy management and conservation engineering. Conducting energy assessments can increase energy system efficiency and reduce energy costs for industrial facilities. Many systems in industrial facilities, including lighting, compressed air, steam, process heating, HVAC, and fans, can benefit from such an energy assessment. There are many opportunities for reducing energy costs for each of these systems, and these opportunities encompass a range of implementation costs. Some opportunities require very little or no capital to implement, while others can extend as far as replacing an entire system requiring a large amount of capital investment to implement.

Author details

Gary P. Moynihan[1*] and Frank L. Barringer[2]

*Address all correspondence to: gmoynihan@eng.ua.edu

1 Department of Civil, Construction and Environmental Engineering, The University of Alabama, Tuscaloosa, AL, USA

2 Department of Mechanical Engineering, The University of Alabama, Tuscaloosa, AL, USA

References

[1] Vanek, F., Albright, L. Energy System Engineering: Evaluation & Implementation. New York, NY: McGraw-Hill; 2008. 532 p.

[2] United States Department of Energy, Energy Efficiency and Renewable Energy. What is the IAC? [Internet]. 2011. Available from: http://iac.rutgers.edu/about.php (Accessed 2016-05-23).

[3] Thurmann, A., Younger, W. Handbook of Energy Audits. 6th ed. Lilburn, GA: Fairmont Press; 2003. 440 p.

[4] United States Department of Energy. 2010 U.S. Lighting Market Characterization. Washington, D.C.: United States Department of Energy; 2012. 100 p.

http://dx.doi.org/10.5772/64902

[5] Murdoch, J. Illumination Engineering—From Edison's Lamp to the Laser. New York, NY: Macmillan; 1986. 541 p.

[6] Rea, M. The IESNA Lighting Handbook: Reference & Application. 9th ed. New York, NY: Illuminating Engineering Society of North America; 2000. 1004 p.

[7] Bloom, S. Induction lighting goes mainstream. Electrical Construction and Maintenance. 2010; 109: 8–14.

[8] Illuminating Engineer Society. IES Footcandle Recommendations. [Internet]. 2013. Available from: http://www.ncat.org/energy/images/IES%20footcandles.pdf [Accessed 2016-06-03].

[9] Harrold, R., Mennie, D. IESNA Lighting Ready Reference. New York, NY: Illuminating Engineering Society of North America; 2003. 205 p.

[10] Barringer, F. Industrial Energy Assessments [thesis]. Tuscaloosa, AL: University of Alabama; 2013.

[11] United States Department of Energy. Improving Compressed Air System Performance: A Sourcebook for Industry. Washington, D.C.: United States Department of Energy; 2003. 128 p.

[12] Brown, R.N. Compressors: Selection and Sizing. 3rd ed. Amsterdam: Elsevier; 2005. 640 p.

[13] American Society of Mechanical Engineers. Energy Assessment for Compressed Air Systems. New York, NY: The American Society of Mechanical Engineers; 2010. 48 p.

[14] Reddy, C., Naidu, S., Rangaiah, G. Waste heat recovery methods and technologies. Chemical Engineering. 2013; 120: 28–38.

[15] Li, Y., Liu, M., Zhang, B. VFD application in a manufacturing plant. Energy Engineering. 2013; 111: 58–77.

[16] United States Department of Energy. Improving Steam System Performance: A Sourcebook for Industry. Washington, D.C.: United States Department of Energy; 2012. 64 p.

[17] American Society of Mechanical Engineers. Energy Assessment for Steam Systems. New York, NY: The American Society of Mechanical Engineers; 2009. 120 p.

[18] Thurmann, A. Mehta, D. Handbook of Energy Engineering. 5th ed. Lilburn, GA: Fairmont Press; 2001. 479 p.

[19] Cengel, Y., Boles, M. Thermodynamics: An Engineering Approach. New York, NY: McGraw-Hill; 2002. 930 p.

[20] United States Department of Energy. Improving Process Heating Performance: A Sourcebook for Industry. Washington, D.C.: United States Department of Energy; 2007. 116 p.

[21] American Society of Mechanical Engineers. Energy Assessment for Process Heating Systems. New York, NY: The American Society of Mechanical Engineers; 2010. 201 p.

[22] United States Energy Information Administration. Commercial Buildings Electricity Consumption by End Use. [Internet]. 2012. Available from: http://www.eia.gov/totalenergy/data/annual/showtext.cfm?t=ptb0211 [Accessed 2016-06-08].

[23] Kavanaugh, S. HVAC Simplified. Atlanta: American Society of Heating, Refrigerating and Air-Conditioning Engineers; 2006. 248 p.

[24] American Society of Heating, Refrigerating and Air-Conditioning Engineers Procedures for Commercial Building Energy Audits. Atlanta: American Society of Heating, Refrigerating and Air-Conditioning Engineers; 2011. 162 p.

[25] Air Movement and Control Association International. Field Performance Measurement of Fan Systems. Arlington Heights, IL: Air Movement and Control Association International; 2007. 69 p.

[26] United States Department of Energy. Improving Pumping System Performance: A Sourcebook for Industry. Washington, D.C.: United States Department of Energy; 2006. 122 p.

[27] American Society of Mechanical Engineers. Energy Assessment for Pumping Systems. New York, NY: The American Society of Mechanical Engineers; 2010; 40 p.

[28] Barnes, M. Practical Variable Speed Drives and Power Electronics. Amsterdam: Elsevier; 2003. 304 p.

[29] Karassik, I., McGuire, T. Centrifugal Pumps. 2nd ed. New York, NY: International Thomson Publishing; 1998. 780 p.

[30] United States Department of Energy. Improving Fan System Performance: A Sourcebook for Industry. Washington, D.C.: United States Department of Energy; 2003. 92 p.

[31] Jorgensen, R. Fan Engineering. 9th ed. Buffalo: Howden Buffalo; 1999. 1072 p.

[32] Owen, M.S. 2004 ASHRAE Handbook, Heating, Ventilating, Air-Conditioning Systems and Equipment (Inch-Pound ed.). Atlanta: American Society of Heating, Refrigerating and Air-Conditioning Engineers; 2004. 888 p.

[33] United States Department of Energy. Improving Motor and Drive System Performance: A Sourcebook for Industry. Washington, D.C.: United States Department of Energy; 2008. 96 p.

Chapter 7

Energy Efficiency of Hydronic Heating System in Retrofitted Buildings

Matjaž Prek

Additional information is available at the end of the chapter

http://dx.doi.org/10.5772/66364

Abstract

Since central-based heating systems with radiators are among the most widely used heating systems in Europe, retrofit measures should include heat distribution and heat emission elements. One of the most cost-effective measures for heating systems is the replacement of the heat generator and circulating pump while preserving the distribution system and heat emitters. Reducing the heating demand results in an oversized existing heating system and thus enables a reduction in flow rates and supply temperatures. These steps should be taken to enhance the efficiency of the heating system without affecting the level of thermal comfort. This chapter focuses on the issue of energy efficiency in retrofitted buildings by optimizing the existing heating systems. The heating equipment is considered as one system, and it is not intended to improve only individual component efficiency. The optimization goal is to achieve a recommended or prescribed thermal comfort level with minimal energy use.

Keywords: energy performance, energy efficiency, hydronic heating, heat emission, thermal comfort, retrofit analysis

1. Introduction

In Europe, the building sector is responsible for 40% of the region's energy consumption and 36% of its CO_2 emissions. Because of high-energy saving potentials, the building sector has become one of the priority areas to meet the EU's targets for 2020 and 2050 [1]. The Energy Performance of Buildings Directive [2] promotes nearly net-zero energy buildings as a mandatory regulation within 2020. This means that new buildings need less than 30–50 kWh/m^2 per year, while existing buildings consume approximately 250 kWh/m^2 per year. Therefore, most of the energy consumption is attributable to the existing building stock. Energy savings obtained from retrofitting existing buildings are more significant than the ones that can be obtained with new buildings.

This fact is reflected in the Energy Efficiency Directive [3] through the requirements for:

1. Public sector to renovate 3% of the central government building stock annually to high-energy performance level (nearly net-zero energy buildings).

2. Energy companies to reduce energy sales by 1.5% every year among their consumers.

Most of the energy savings of the retrofit can be contributed by two key targets: reducing heating demand and increasing the efficiency of HVAC systems. An important factor in achieving the expected reduction target is improving the efficiency of systems without affecting thermal comfort. Typical retrofit measures to reduce the heating demand include applying insulation layers on walls and/or replacing windows. Little attention is given to how the improvements on the building's envelope affect the settings and efficiency of the heating system. Basically the increase of insulation reduces the heating demand which in turn can reduce flow rates and/or supply temperatures. Reducing the supply temperature makes low-temperature heating system interesting and economically viable [4, 5].

Since central-based heating systems with radiators are among the most widely used heating systems in Europe [6], retrofit measures should include heat distribution and heat emission elements. One of the most cost-effective measures on heating systems is the replacement of the heat generator and circulating pump while preserving the distribution system and heat emitters [7]. Reducing the heating demand results in an oversized existing heating system and thus enables a reduction in flow rates and supply temperatures. These steps should be taken to enhance the efficiency of heating system without affecting the level of thermal comfort.

This chapter focuses on the issue of energy efficiency in retrofitted buildings by optimizing the existing heating systems. The heating equipment is considered holistically and it is not intended to improve only individual component efficiency. The following steps are presented:

- The need for optimization
- Impact of heating technology on the optimization process
- Optimization of the hydraulic network

An integrated methodology for energy efficiency in retrofitted buildings is proposed through coupling active and passive strategies specifically tailored for application in traditional heating systems. The main results show how the applications of specific building energy retrofit actions could increase the energy efficiency of the heating system without compromising thermal comfort. Finally, a system optimization is also determined by different constraints, i.e., the use, economy rules, and technical regulations. Nevertheless, the optimization and improvement of the heating system efficiency in retrofitted buildings give a unique opportunity that new buildings are not given, namely, the possibility of energy saving [8].

http://dx.doi.org/10.5772/66364

2. Building envelope refurbishment

Contemporary type of building construction with smaller heat losses and demands for increased comfort requires new concept of radiator construction and regulation of heat output. The heat emission of a radiator depends on size, design, and mean temperature difference between the radiator surface and the indoor temperature—excess temperature. At given thermal constraints, the effect of geometrical and thermal parameters of a radiator could be optimized [9]. On the other side, to maintain the desired heat output, it must be regulated according to the condition of thermal comfort and outdoor conditions.

Control of the heat output can be realized by changing the radiator surface, by changing the heating water inlet temperature (temperature control), or by changing the mass flow through the radiator (mass control) [10]. The first option is not physically feasible; another option is feasible using relatively complex control system with three-way mixing valve. This control method is used primarily in the central control unit, which regulates the temperature of heating water to a large number of radiators. For local regulation of the heat output, the most commonly used principle is by varying the mass flow of water through the heater, where the flow change is controlled by a throttle valve. This system is easy and most affordable, so almost all systems of local regulation of radiators are based on "throttling." From the standpoint of regulation, this system has several shortcomings. Since the radiator heat output varies exponentially according to the mass flow rate, in the case of large throttling, a small change in valve position (small change in flow rate) causes a disproportionately big change in heat output. This situation occurs at reduced heat output, oversized radiator (also as a result of proper dimensioning taking into account the heating-up reserve for intermittent/reduced mode of operation of heating system), inappropriate (too high) water inlet temperature, or because of internal and external heat sources.

Another problem is the heat output at reduced water flow rate [11]. Due to the smaller water speed in the radiator, the retention time is longer, resulting in a lower water temperature at the exit of the radiator and a lower average temperature of the radiator. Low water flow rate also causes mixing of inlet water with colder water in the radiator. These result in a less effective inlet water temperature, so the heat emitted is significantly different—lower—than expected, according to standard calculation methods with respect to excess temperature. Heating systems mostly operate at variable loads, which depend on the regulation. One of the tasks of heating systems is not only to provide the necessary heat but also to adapt the heat output as quickly as possible to the change of heat load. From this viewpoint, it is recommended to have heating systems (including radiators) with low thermal inertia.

The refurbishment of the building envelope affects the thermal characteristics, which have a significant effect on the heating system operating conditions and consequently on its energy efficiency. To ensure the latter, we must consider the following aspects: the efficiency of each component and the efficiency of the system, which is determined with the interactions of individual components. For the whole system efficiency, the following conditions must be met:

- properly sized and set system elements (radiators, valves, piping system, pumps, heat generators)
- setting of local and central regulation
- optimal hydraulics

These conditions apply both to new and existing buildings. Most of the heating system components are already determined in existing buildings. The refurbishment of the building envelope therefore requires an adjustment of existing system components to achieve optimal operation and consequently proper energy efficiency. The fundamental difference between the optimization of an existing and new system is that optimal operation of existing systems is achieved with adjusting their settings, whereas the optimization of new systems is based on selecting components of proper quality and characteristics.

Optimization and efficiency are closely connected with internal environment quality and operating conditions. Internal environment conditions that enable a proper quality of living or a proper thermal comfort level must be considered [12]. Thermal comfort level of a certain thermal environment directly determines energy use. Therefore, we cannot discuss the heating system efficiency without the achieved thermal environment. The optimization goal is to achieve a proper thermal comfort level with minimal energy use.

The refurbishment of the building envelope results in a decrease of the required heat for heating and the power of radiators. Therefore, operation of the heating system has to be adjusted to the new operating conditions. After the building envelope refurbishment, the heating load of the outer rooms generally decreases more significantly than at the inner rooms. The heating load decrease depends on the improvement of the building envelope thermal transmittance and the fraction of outer wall area. A larger fraction of outer wall area results in a more significant decrease in the required radiator power. However, thermal insulation does not cause a significant decrease in the required radiator power for inner rooms. Consequently, it is important that we after the building envelope refurbishment properly adjust the thermal power of radiators according to the heating load decrease in the room in which they are installed.

3. Radiator heat output

The radiator thermal power is defined by the supply temperature and the mass flow. For the determination of the required thermal power of the radiator heat, losses due to transmission (Eq. (1)) must be considered:

$$\dot{Q}_T = U_e \cdot A_e \cdot (t_i - t_e) \tag{1}$$

where $\dot{Q}_T$ is the heat losses due to transmission; U_e is the thermal transmittance; A_e is the area of external wall; t_i is the internal temperature; t_e is the external design temperature.

The change of the heat load before and after the envelope refurbishment is therefore:

$$\frac{\dot{Q}_{T1}}{\dot{Q}_{T2}} = \frac{U_{e1} \cdot A_e \cdot (t_i - t_e)}{U_{e2} \cdot A_e \cdot (t_i - t_e)} \tag{2}$$

where index 1 designates the properties before the refurbishment and index 2 the properties after the latter. It is assumed that the areas of external walls as well as internal and external temperature do not change.

The required heat is provided by the radiator, whereby its thermal power is defined according to Ref. [13]:

$$\dot{Q} = A \cdot \Delta t_{ex}^{n} \tag{3}$$

where $\dot{Q}$ is the thermal power of radiator; A is the radiator area; Δt_{ex}^{n} is the excess temperature; N is the radiator exponent.

The radiator excess temperature is defined as a logarithmic temperature difference between the average radiator surface temperature and the room temperature:

$$\Delta t_{ex} = \frac{t_S - t_R}{ln \frac{t_S - t_i}{t_R - t_i}} \tag{4}$$

where t_S is the supply temperature; t_R is the return temperature; t_i is the room temperature.

The change of the emitted heat by the radiator is defined with the following relation:

$$\frac{\dot{Q}_1}{\dot{Q}_2} = \frac{A_1 \cdot \Delta t_{ex1}^{n}}{A_2 \cdot \Delta t_{ex2}^{n}} \tag{5}$$

where A_1 is the radiator area before the refurbishment; A_2 is the radiator area after the refurbishment; Δt_{ex1}^{n} – is the excess temperature before the refurbishment; Δt_{ex2}^{n} – is the excess temperature after the refurbishment.

If we assume that the radiator area does not change and that the radiator exponent is constant, the change of the emitted heat depends only on the excess temperature change:

$$\frac{\dot{Q}_1}{\dot{Q}_2} = \left(\frac{\Delta t_{ex1}}{\Delta t_{ex2}}\right)^{n} \tag{6}$$

The heat required for heating is provided with the flow of heated water through the radiator. The heat emitted from water depends on the mass flow and cooling down of the water:

$$\dot{Q} = \dot{m} \cdot c_p \cdot (t_S - t_R) \tag{7}$$

where $\dot{Q}$ is the radiator thermal power; $\dot{m}$ is the water mass flow through the radiator; c_p is the water specific heat; t_S is the supply temperature; t_R is the return temperature.

The change of the emitted heat is so defined with the following relation:

$$\frac{\dot{Q}_1}{\dot{Q}_2} = \frac{\dot{m}_1}{\dot{m}_2} \cdot \frac{(t_{S1} - t_{R1})}{(t_{S2} - t_{R2})} \tag{8}$$

where index 1 designates the state before while index 2 designates the state after the refurbishment.

3.1. Radiator heat output at various conditions

From Eqs. (2), (6), and (8), the influence of building thermal envelope change on the change of radiator heat output can be seen. It can also be seen that this change can be compensated with:

- Changing the water supply temperature (Eq. (6))
- Changing the water mass flow (Eq. (8))

Hereby, we considered the interdependence between Eqs. (6) and (8). At constant water temperature, the return and excess temperatures change because of the water mass flow change. The interdependence between water mass flow and the radiator excess temperature can be depicted graphically [14, 15]. **Figure 1** shows the emitted heat of a radiator with radiator exponent n = 1.3 at standard temperature conditions: supply temperature t_S = 75°C, return temperature t_R = 65°C, and room temperature t_i = 20°C.

The emitted heat is shown as a relation between the heat output at standard conditions $\dot{Q}_0$ and at other conditions $\dot{Q}$. The supply and return temperatures are included with the supply and return excess temperature and the dependence between heat output and mass flow is expressed as a ratio between the actual and standard mass flow. Thereby, all the influences of the heating system on the heat output are considered. With such a diagram (or numerically with an iterative procedure), one can determine the new operating conditions for an existing radiator.

Figure 2 shows an example of the effect of heat loss reduction on the change of radiator operating conditions.

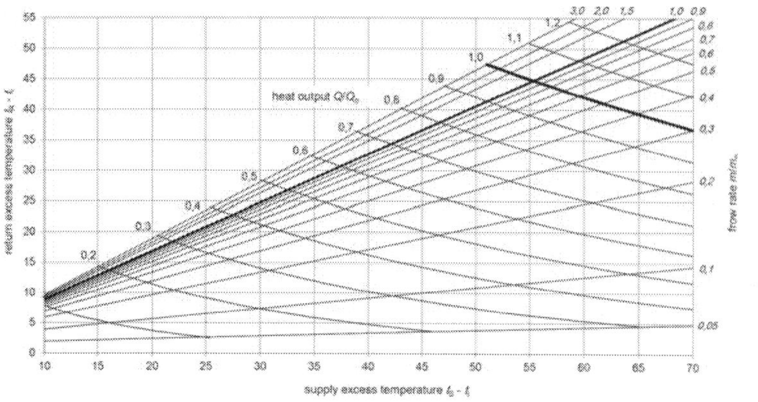

Figure 1. Heat output of the radiator at operating conditions 75/65/20 [13, 15].

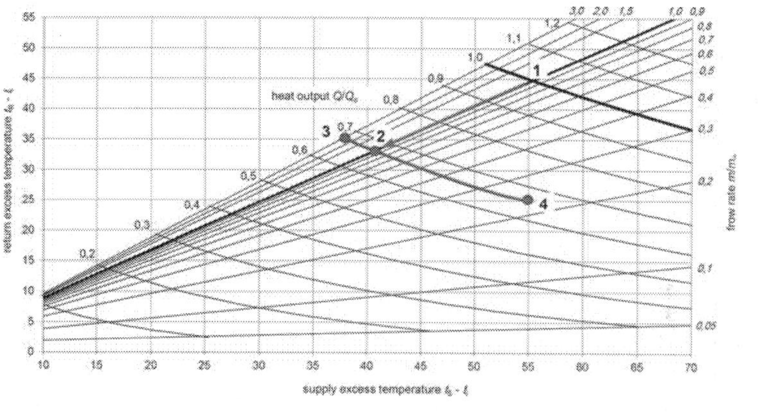

Figure 2. Heat output at the reduced heat demand for 33%.

Point 1 represents the radiator thermal power before the refurbishment, whereby the temperature regime 75/65/20 [13] was assumed. If the radiator is properly sized and selected, the thermal power is $\dot{Q}/\dot{Q}_0 = 1.0$, the mass flow is $\dot{m}/\dot{m}_0 = 1.0$, and the logarithmic excess temperature is 49.83 K. The refurbishment of the building envelope results in a decrease in the required radiator thermal power, e.g., 33%. Therefore, the reduced heat output of the radiator is $\dot{Q}/\dot{Q}_0 = 0.67$. Maintaining the (unchanged) mass flow $\dot{m}/\dot{m}_0 = 1.0$ (blue line), we reach the required heat output at the mean logarithmic excess temperature of the radiator 37 K—point 2. However, the radiator reaches the same heat output at different supply and return water temperature combinations. Possible combinations are shown as a red line in **Figure 2**; characteristic points on this line represent the following operating conditions:

—Point 2: mass flow $\dot{m}/\dot{m}_0 = 1.0$, supply temperature $t_S = 62°C$, return temperature $t_R = 53°C$.

—Point 3: mass flow $\dot{m}/\dot{m}_0 = 3.0$, supply temperature $t_S = 57°C$, return temperature $t_R = yy°C$.

—Point 4: mass flow $\dot{m}/\dot{m}_0 = 0.24$, supply temperature $t_S = 75°C$, return temperature $t_R = 45°C$.

The difficulty in determining the heating system operating parameters is that the reduction in the required thermal power is generally not the same for the other rooms (other radiators). For a room with a higher fraction of external surfaces, the reduction in the required thermal power is more significant, e.g., 40%. This case is depicted in **Figure 3**.

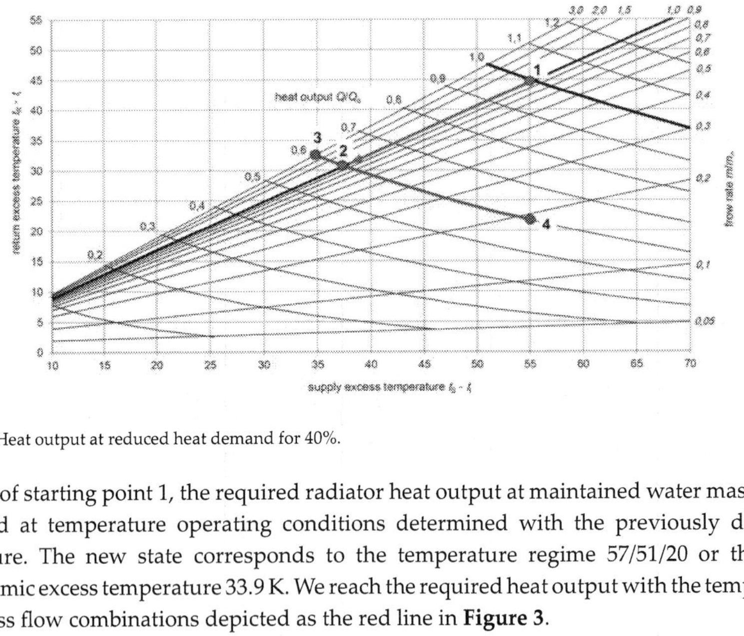

Figure 3. Heat output at reduced heat demand for 40%.

Instead of starting point 1, the required radiator heat output at maintained water mass flow is achieved at temperature operating conditions determined with the previously described procedure. The new state corresponds to the temperature regime 57/51/20 or the mean logarithmic excess temperature 33.9 K. We reach the required heat output with the temperature and mass flow combinations depicted as the red line in **Figure 3**.

Figures 2 and **3** show that a higher flow rate can be necessary despite the lower thermal power of the radiator after the refurbishment (point 3 in **Figures 2** and **3**) or a supply temperature equal or lower to the previous value (point 4 in **Figures 2** and **3**). The dependency between the flow rate and temperature is depicted with a red line between points 3 and 4. Therefore, in the first step, we can determine the boundary conditions, which define optimal operation of the heating system:

- The water supply temperature should be lower or equal to the value before the refurbishment.
- Water flow rate should be lower or equal to the value before the refurbishment.

We can depict the restrictions on the operating parameters of the heating system as constrictions of individual radiator operating parameters. **Figure 4** shows the operating parameters of both radiators. The maximal supply temperature is limited with point 1, whereas the minimal value is limited with point 2.

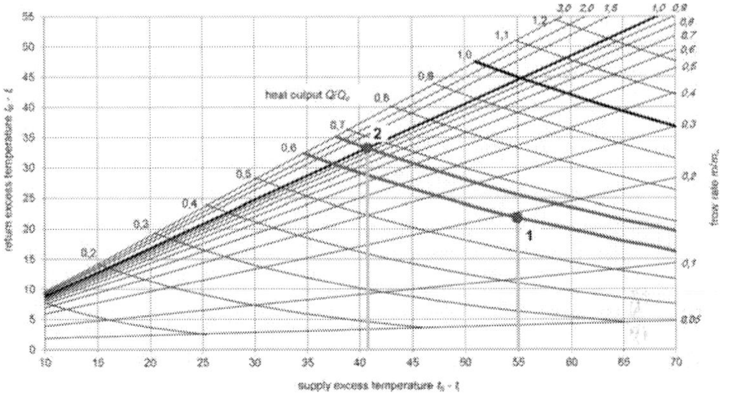

Figure 4. Operating conditions for both radiators after refurbishment.

Since the supply temperature must be equal for all radiators, an intermediate temperature is chosen, e.g., t_S = 65°C. In this case, shown in **Figure 5**, the mass flow for both radiators is determined: $\dot{m}_1/\dot{m}_0 = 0.37$ and $\dot{m}_2/\dot{m}_0 = 0.52$.

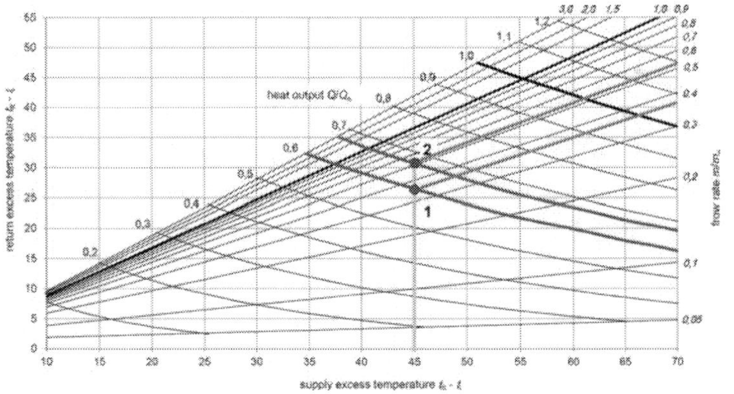

Figure 5. Radiator heat output at a unified supply temperature.

With the heating system temperature, we determine the type of the heat generator or its temperature regime. If we want to use a gas-fired condensing boiler or a heat pump, the required maximal supply temperature is t_S = 55°C. This temperature represents the threshold temperature of the heating system with which we achieve thermal comfort conditions. For heating systems with a lower temperature, a replacement of radiators with other heating element types (fan-coil radiators, fan-coil units), where the heat output does not depend on natural convection, is therefore applicable.

However, in the described example, a supply temperature decrease is possible at the assumption that the critical radiator is replaced. **Figure 6** depicts the case in which the system supply temperature in additionally lowered to t_S = 60°C (yellow line shift on **Figure 6**). The radiators heat output is maintained.

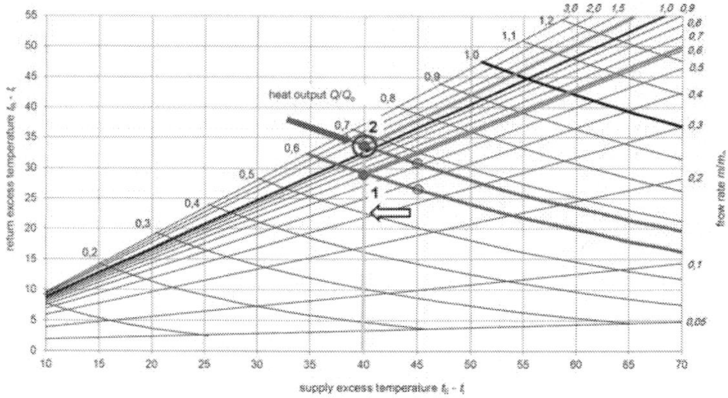

Figure 6. Influence of reduced supply temperature.

Because of the lower supply temperature, a higher water mass flow is required. The required flow rate for radiator 1 is $\dot{m}_1/\dot{m}_0$ = 0.61, which is still lower than the nominal flow rate. For radiator 2 a flow rate of $\dot{m}_2/\dot{m}_0$ = 1.2 is required, which exceeds the base flow rate. We can avoid a higher flow rate in the heating system by replacing the critical radiator with a radiator with a higher nominal heat output.

4. Conclusions

During building retrofit energy use efficiency and cost-effectiveness must also be considered besides the costs. In practice, the retrofit of a building usually consists of a building envelope refurbishment and heat generator replacement. Because of the high cost and the interference in living areas, the heating system (pipping and radiators) is not significantly altered. The envelope refurbishment results in a decrease in the heat required for heating and consequently in the lowering of the required thermal power of the radiators and the heat generator. The lowered thermal power enables the lowering of the heating system temperature regime and thereby the option of using a more efficient heat generator. However, heating system efficiency is conditioned with ensuring thermal comfort, which must be met in every case.

In this work, a method for determining the heating system operation conditions is presented. The problem of uneven reduction in the required thermal power of radiators is exposed, which is a consequence of uneven decreases in room heat losses. A larger fraction of outer wall area results in a more significant decrease in the required radiator power in that room. Therefore, new operation parameters must be determined based on the load of individual radiators. The presented method is demonstrated on a case of two radiators with different thermal characteristics.

The influence of radiators on the building system's energy efficiency is taken into account in two respects: with the determination of the whole system's temperature regime and the required mass flow rate. Both parameters are interdependent; thus, parameter optimization is achieved under the consideration of proper boundary conditions. In the demonstrated example, the following conditions were considered: the flow rate must be lower or equal to the existing one—hereby minimal energy use of the circulating pump is assured. The second restriction represents the supply temperature, which must be lower or equal to the existing value, which is a precondition for efficient heat generator operation. The temperature of radiators required to still ensure thermal comfort of the room was also considered. The combination of both parameters ensures optimal and energy-efficient operation of the heating system.

Author details

Matjaž Prek

Address all correspondence to: matjaz.prek@fs.uni-lj.si

Faculty of Mechanical Engineering, University of Ljubljana, Ljubljana, Slovenia

References

[1] European Commission. Energy Efficiency Plan. COM(2011) 109 final. Tech. rep. Communication. 2011

[2] European Union. Directive 2010/31/EU of European Parliament and of the Council of 19 May 2010 on the Energy Performance of Buildings (recast). 2010.

[3] European Union. Directive 2012/27/EU of the European Parliament and of the Council of 25 October 2012 on Energy Efficiency, Amending Directives 2009/125/EC and 2010/30/EU and Repealing Directives 2004/8/EC and 2006/32/EC. 2012.

[4] Tunzi M., Ostergaard D.S., Svendsen S., Boukhanouf R., Cooper E. Method to investigate and plan the application of low temperature district heating to existing hydraulic radiator systems in existing buildings. Energy. 2016;113:413-421.

[5] Ostergaard D.S., Svendsen S. Case study of low-temperature heating in an existing single-family house—A test method for simulating of heating system temperatures. Energy and Buildings. 2016;126:535-544.

[6] Peeters L., Van der Veken J., Hens H., Helsen L., D'haeseleer W. Control of heating systems in residential buildings: current practice. Energy and Buildings. 2008;40:1446-1455.

[7] Penna P., Prada A., Cappelletti F., Gasparella A. Multi-objectives optimization of energy efficiency measures in existing buildings. Energy and Buildings. 2015;95:57-69.

[8] Nagy Z., Rossi D., Hersberger C., Irogoyen S.D., Miller C., Schlueter A. Balancing envelope and heating system parameters for zero emissions retrofit using building sensor data. Applied Energy. 2014;131:56-66.

[9] Arslantürk C., Özgüç A.C. Optimization of a central-heating radiator. Applied Energy. 2006;83(11):1190-1197.

[10] Ward I.C.. Domestic radiators: performance at lower mass flow rates and lower temperature differentials than those specified in standard performance tests Building Services Engineering Research and Technology. 1991;12(3):87-94.

[11] Beck S.M.B., Grinsted S.C., Blakey S.G., Worden K. A novel design for panel radiators. Applied Thermal Engineering. 2004;24:1291-1300.

[12] EN 15251-2007. Indoor Environmental Input Parameters for Design and Assessment of Energy Performance of Buildings Addressing Indoor Air Quality, Thermal Environment, Lighting and Acoustics. Brussels: European Committee for Standardisation. 2007.

[13] EN 442-2: 1996/A2:2003. Radiators and Convectors—Part 2: Test Methods and Rating. Brussels: European Committee for Standardisation. 2003.

[14] Gray F.M., Schmidt M.. Thermal building modelling using Gaussian processes. Energy and Buildings. 2016;119:119-128.

[15] VDI 6030 Part1: 2002. Designing free heating surfaces—Fundamentals—Designing of heating appliances. Berlin: Beuth Verlag. 2002.

Chapter 8

Managing Energy Demand in Buildings through Appropriate Equipment Specification and Use

Abimbola Olukemi Windapo

Additional information is available at the end of the chapter

http://dx.doi.org/10.5772/66363

Abstract

The high demand for electrical energy in virtually all human endeavour has engendered the continuous rapid growth of electricity production, transmission and distribution worldwide. Every habitable building structure usually requires electrical appliances and mechanical systems such that the cost of electrical and mechanical installation in a building is generally between 10% and 35% of the total construction cost. This chapter examines the equipment used for electrical, mechanical and lighting systems in contemporary buildings in Cape Town, South Africa, towards determining those materials and equipment aiding energy efficiency in these buildings. The research employs a multiple case study approach, consisting of recently completed buildings. The study established that the equipment used in these contemporary buildings to effect a reduction in energy consumption are compliant with the main specifications and policies guiding energy efficiency in buildings in South Africa and that owners of the Case Study buildings obtained a significant reduction in power consumption as a result of the installation of the identified equipment. Based on these findings, the study concludes that a building that is compliant with energy efficient systems installation standards will experience a significant reduction in utility bills, and savings for commercial buildings and private property owners.

Keywords: efficiency, energy consumption, energy demand

1. Introduction

Globally, the demand for energy-powered services (energy efficiency and embedded generation) is growing [1]. Modern technology favours the use of electrical energy due to its numerous technical and economic benefits when compared to other forms of energy. The cost of electrical and mechanical installations in buildings is usually in the range of 10–35% of the total new building construction cost [2, 3]. Electrical and mechanical installation consists of electrical services, fire protection, plumbing, vertical transport and heating, ventilating and air-conditioning (HVAC) [3]. The cost of electricity in commercial buildings is very high due to high energy utilisation factor. According to Mobley [4], commercial building owners spend 22.4% of their operating costs on utilities (energy and water). Electricity prices are expected to rise in South Africa until they reflect the costs of the new infrastructure built to generate electricity to deal with the country's energy gap. According to GreenCape [1], the global market for energy services is driven by growing awareness of the impact of carbon emissions, rising electricity costs and increased financial returns arising from energy investments. This chapter examines the equipment used for electrical, mechanical and lighting systems in selected contemporary buildings in Cape Town; how these are specified; power demand and energy consumption in buildings and its relationship to energy bills; and steps for determining equipment that aid energy efficiency. Four case studies consisting of commercial buildings (offices and hotels) located in Cape Town were used as sources of primary data. The method of data collection used in the study include interviews, with building owners, on-site inspections and records (architectural, mechanical, electrical and plumbing drawings) that show how certain equipment and system have been used to advance energy efficiency (generation and conservation) in the buildings.

2. Outline of buildings used in the study

The buildings used in the study are contemporary buildings located in Cape Town. The justification for the use of these buildings are that the building must have been completed within the last 5 years and were designed based on 'green' building principles with the anticipation of attaining a 'green' building rating score. Three of the buildings used in the study were awarded green star ratings on completion. A description of the buildings used in the study is outlined below.

2.1. Case study A

Case study A is a 11,624 m^2 six storey hotel with a seemingly simplex layout, designed in an L-shape. The hotel owner strives to minimise the hotel's carbon footprint and decrease energy utilisation. As a result, this building has many highly advanced innovative systems including 3 kW wind turbines, photovoltaic panels, vegetated green roof, geothermal loops, occupancy sensors and regenerative lifts, imported from all over the world and installed to minimise the day-to-day running costs of the hotel.

2.2. Case study B

Case study B is the new Head Office building for an oil and gas company. The building situated in Cape Town, achieved the first 'green' certification in the industry category of refiners and marketers of petroleum products in South Africa. The building is a 9000 m^2 five levels (1 basement, ground, mezzanine, 1st and 2nd floor) in a square form. The building is a framed structure comprising of a large central courtyard. The project team sought to achieve energy conservation through the reduction of reliance on nonrenewable energy sources and the necessary criteria of the Green Building Council of South Africa. A fully integrated DALI lighting system, complete mechanical ventilation to all usable areas and the basement and a sub-metering system are equipment used in the building to enable a reduction in energy consumption.

2.3. Case study C

Case study C is a newly developed office building located in Cape Town. The building is an 18,600 m^2 office building, with approximately 500 m^2 retail and 455 car parks extending over two basements levels. The building is a ground breaking project in the South African context as it is the only 6-Star Green Star SA-Design and as Built certified building in South Africa. The idea for the building came about and was driven by the unique combination of the developer and tenant. This chapter will discuss the initiatives such as a Sea Water Cooling System, an ICT system that monitors energy and water use, a lighting system that will ensure that lighting is only used when required and a green roof that contributes to thermal roof insulation, implemented in order to obtain Green Star points, as well as detail how they operate in order to mitigate the carbon footprint of the building.

2.4. Case study D

Case study D is a 124 m high, 34-floor multi-story office building located in a prime location on the Cape Town foreshore. The building has a footprint of 114,547 m^2 comprising of office space, seven parking levels, an atrium and a Sky plaza. The building was awarded a 5-Star Design rating under the Green Building Council of South Africa's Green Star rating system, making it the first 5-Star Green Star skyscraper in the country. A significant amount of engineering, planning and innovative forward thinking had to be carried out by the professional team. The professional team has focussed on energy reduction initiatives, through the adoption of an efficient air-conditioning system and application of LED lighting fittings throughout the office space.

3. Overview of equipment used for electrical, mechanical and lighting systems in the case study buildings

The public power generation company in South Africa Eskom is the main supplier of electricity to the buildings studied. South Africa is in the midst of an energy crisis that is resulting in rising energy prices, rolling blackouts (known as load shedding in South Africa) and changing energy policies and incentives [1]. Power outage/load shedding comes with significant costs to businesses that are unable to operate during periods of power outage shutting down production at companies [5], and causing inconveniences to residents [6]. As a result of this, end users in the commercial building sector (offices and hotels) have begun exploring alternative energy options through the use of appropriate equipment [1]. The case study buildings generated additional power from solar panels, wind turbines and a regenerative lift. Electrical appliances and mechanical systems are required in buildings to make the building habitable. Initiatives such as sea water cooling air-conditioning units, displacement ventilation units, geothermal loop systems, LED lighting, automated lightning and electronics that automatically turns off when rooms are unoccupied (occupancy sensors) were implemented to reduce operational costs. The power-generating equipment together with other energy saving techniques is said to reduce the consumption of electricity from the mains by 35–94% each year (field survey). In the event of a power outage, electrical energy from the batteries can be used to run buildings for at least 30 min (field survey–case study A), which gives the generator time to start.

3.1. Electricity generation equipment

The photovoltaic solar panels, wind turbine and regenerative lifts discussed in this section are used as alternative means of electricity generation in case study A (hotel building) to reduce the reliance of the hotel on public power supply. These equipment are examples of alternative forms of electricity generation, which make it possible to reduce energy demand in buildings and increase energy efficiency.

3.1.1. Photovoltaic solar panels

Case study A makes use of a number of solar panels to generate electricity. The 240 W photovoltaic solar panels, as shown in **Figure 1**, are produced locally in South Africa and have an anticipated production of 78,000 kWh/year. The panels are connected in series and positioned on the buildings to face the North-East in order to ensure that the panels maximise the sun energy. The system includes an Afrisum Karoo 70 KVA Grid tied inverter as shown in **Figure 2**, installed to serve as the core of the solar energy installation system, as it harvests the energy captured by the solar panels, and a battery bank for energy storage, and a charge controller that regulates the power flow into and out of the battery bank. Battery banks are typically sized in order to provide energy during days of no or limited sunshine.

Figure 1. 240 W photovoltaic solar panel.

http://dx.doi.org/10.5772/66363

Figure 2. Afrisum Karoo 70 KVA Grid Tied Inverter.

3.1.2. Huebner-Giessen 3 Kw wind turbine

Three vertical axis wind turbine that convert wind into electricity were installed in front of case study building A at a height of 17 m to maximise output as shown in **Figures 3** and **4**. Cape Town is well known for the high amount of wind that it generates. Therefore, using wind turbines as a power generating system generates power without using fossil fuels or producing gasses or toxic waste. Location is critical for the use of wind turbines since buildings can obstruct the wind flow. These turbines use the wind, a renewable energy source, to contribute 9 kWh to the hotel by converting wind energy into electrical energy. Although the wind turbine does not generate a lot of energy, it makes a small change and creates social awareness about alternative sources of energy.

Figure 3. Huebner-Giessen 3 kW wind turbine here.

Figure 4. Huebner-Giessen 3 kW wind turbines.

3.1.3. Regenerative elevators

An elevator can account for 5–10% of the energy consumption of a building. In case study A, there are regenerative energy drives on all elevators. Energy is saved by installing energy-efficient lifts fitted with regenerative drives and brake system which recapture 30% of the input energy and feed it back into the building. This follows the principles that in order to move an elevator, energy must be produced to accelerate the lift. Likewise, energy must be removed in order to decelerate the lift as shown in **Figure 5**. Twin Kone elevators fitted with a regenerative braking system were installed in case study A.

When using a mechanical brake, the removed energy is wasted in heat. A cooler will therefore be required for the machine room. By contrast, a regenerative motor will recover potential energy stored from an elevator's use and feed it back to the unit as electricity. The elevator utilises this surplus energy and transfers it back into the building's electrical system for use in other areas. More specifically, the regenerated energy is connected to a battery storage bank that houses large batteries and charges the batteries as the lift brakes.

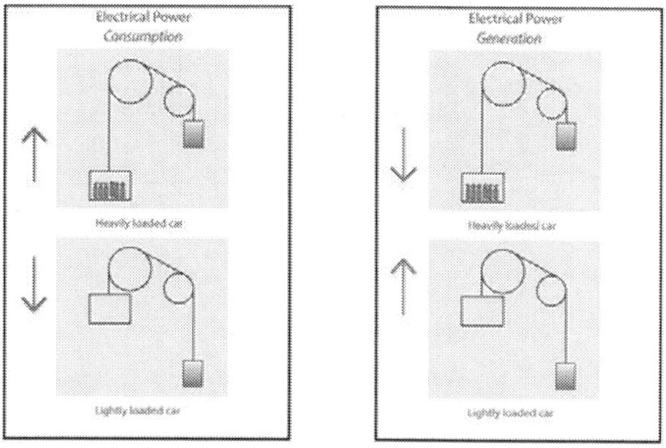

Figure 5. Electricity consumption versus generation.

The gearless traction machine, control panel and vector-controlled system are installed within the hoist pathway located on the side wall of the lift shaft as shown in **Figure 6**. This arrangement of equipment frees up space normally required for separate machine rooms. The use of regenerative lift technology targets significant savings in energy costs, financial and reduction in environmental impact.

Energy received from the mains, solar panel, wind turbines and regenerative drive lift installed in case study A is stored in a battery room as shown in **Figure 7**.

3.2. Typical mechanical equipment used in the selected buildings for energy conservation

Mechanical equipment used in the buildings, apart from the regenerative lift shown in **Figures 5** and **6**, include a sea water cooling system used in case study C and mechanical ventilation systems used in case studies B, C and D. These are further described in this section.

3.2.1. Sea water cooling system

A sea water cooling system in which cold sea water is used for cooling units of the air-conditioning plants as shown in **Figure 8** was installed in the basement of case study building C. Unlike ordinary water-cooled air-conditioning systems, which make use of cold potable water, this air-conditioning system makes use of cold Atlantic seawater drawn from the harbour, where the water is normally between 14 and 16°C through a titanium plate heat exchanger [7]. This cooled water is then used to cool down the chiller plant when it rejects heat from the building. The sea water cooling plant was a radical intervention used by members of the project team in solving the problem of energy conservation when cooling the building.

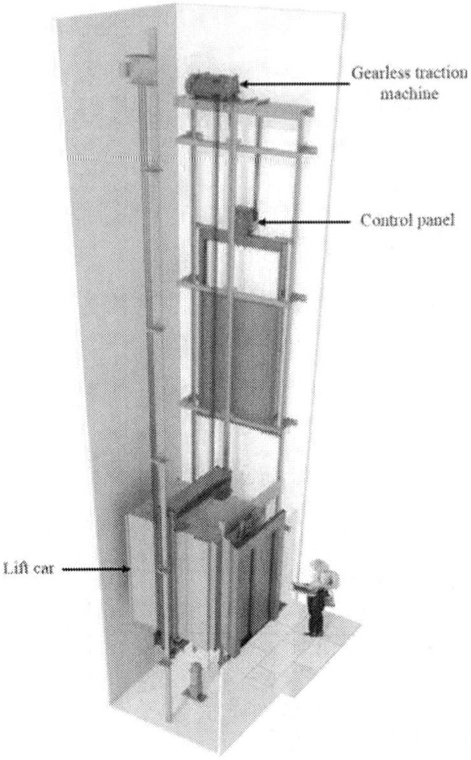

Figure 6. Traction less elevator shaft detail.

Figure 7. The batteries used to store generated energy.

http://dx.doi.org/10.5772/66363

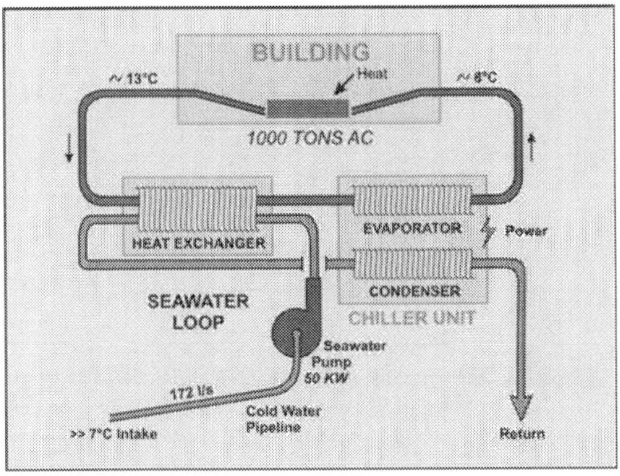

Figure 8. Diagram of sea water cooling system [8].

3.2.2. Displacement ventilation

Case study building C makes use of air handling equipment placed at intervals on the roofs as shown in **Figure 9**. A mechanical inlet and mechanical extract system that provides air at slow velocities as opposed to the usual high speed velocities are used to regulate and balance the supply and emission of air. The air in a room is circulated based on the following process—cool air is sent through a diffuser, the cool air is warmed by body heat and rises, creating an upward convection that draws fresh air across individuals, the warmer used air and pollutants are extracted at ceiling level, the air handling equipment recovers wasted energy, wrings out moisture and slightly cools fresh air, which is then delivered into the building through a diffuser. Due to this innovative ventilation system, it is found that the internal environmental air quality within these buildings is very good and reportedly 150% better than the legislated standard [9].

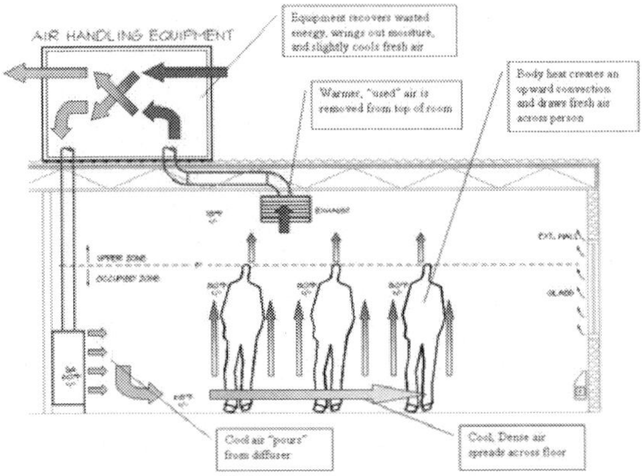

Figure 9. Displacement ventilation [10].

The use of a ventilation system that ensures sufficient supply of fresh air is used in case study B. Economy cycle technology is used for the air-conditioning fans, which determines the quantity of fresh air required and then also recycles a quantity of cooler return air, which reduces the energy used in cooling the air. Variable speed fans are used in delivering air to the building at the rate needed to circulate the conditioned air. Since case study B is long both on the East and West side, at certain times during the day, the building experiences both heat loss and gain at the same time in different areas. The HVAC system will be notified of this through the Building Management System (BMS) and adjusted through building heat produced by chillers acting as heat pumps, thereby eliminating the need for less energy-efficient electric heating.

3.2.3. *Geothermal loop system*

The geothermal loop system installed under the foundation of case study A, comprise of four Alpha–Innotec ground source heat pumps, coupled to 100 boreholes each 76 m deep, into which high density polyethylene (HDPE) U-bend pipes are installed. Subsequently heating or cooling water would be circulated through these pipes. These pipes would be connected together using headers and linked to the ground source heat pumps which are the central system of case study A's heating ventilating and air-conditioning (HVAC) system. The geothermal loop system was imported from Germany since it is not available locally. Air-conditioning is a huge energy consumer for hotels and the geothermal system is alleged to reduce the energy consumption of case study A by about 25%. This system uses water to keep the ground temperature below case study A at a constant 19°C throughout the year and significantly reduce the energy used for heating and cooling water and also the energy consumed by air-conditioning equipment. The 100 holes and all the high density piping used for the geothermal loop installation were positioned with protective foam to prevent any damage to the pipes. The piping and holes are covered by the basement concrete surface bed which keeps all the piping protected and out of sight. The only visible equipment as shown in **Figure 10** is the 14 sub-header pipes that penetrate through the concrete surface bed and connected to the heat pumps which feed the HVAC system.

Figure 10. Geothermal loop system hole connections with valves.

http://dx.doi.org/10.5772/66363

3.3. Common lighting system installed in the selected buildings

Digital Addressable Lighting Interface (DALI) lighting systems: Sophisticated lighting systems known as DALI are installed in case study B. The DALI lighting system is a two-way communication system that connects digital technology to lighting. The system allows the lights to communicate to the user, and allows the user to communicate back to the system using DALI regulators as shown in **Figure 11**, computers equipped with appropriate software or the appropriate Building Management System (BMS). This system allows the occupant to regulate the lighting with each light being individually addressable, which means each light can be dimmed or switched off depending on the users' preference. The DALI lighting system has sensors, which have a combination of being manually switched off and on and infrared readers which can sense if there are people in the building and automatically switches off the light if an area is not used.

Figure 11. Example of a Digital Addressable Lighting Interface (DALI) regulator.

Dali lighting systems have the ability to sense daylight in a specific area in the building and can automatically dim lights using artificial lighting and daylight to preserve luminance levels to reduce energy usage: this process is known as daylight harvesting. The lights are controlled by heavy materials such as gravel, sand or iron, which have high frequency regulators and produce no flicker or white noise. The use of these heavy materials in DALI lighting systems prevents the occupants from suffering eyestrain and headaches, which are caused by the older magnetic ballast installation lights. The additional benefit is a 50% reduction in energy losses due to ballast installation.

The lamps used by the DALI system are known as T5's (high frequency small diameter fluorescent lamps). It is acknowledged that these lights provide the highest industry standards in terms of lighting quality, colour, energy efficiency, with reduced maintenance costs and an average life span of 4 years per fluorescent. The lights can be dimmed to 10% of its total capacity when necessary thereby providing improved energy savings.

Occupancy/motion sensors for lighting were also installed in case study A, B, C and D in all public areas and stores. This system ensures that lighting is only used when required. When the space is not in use, the lights are automatically turned off and signalisation displays are dimmed. In addition, LED lights that use up to 80% less energy and last longer than conventional lighting is used in all the buildings.

3.4. Smart systems used in the selected buildings

Electricity sub-metering: An electricity sub-metering system is installed in case study B, in compliance with the proposed Smart Metering Standards (NRS 049) of South Africa. Electricity sub-metering measures energy consumption and provides users with the data necessary to monitor and control energy use in buildings. For example, if an area in a building is using a lot more energy for lighting than other areas, the building operator can identify this with the use of sub-metering and investigate the cause. Sub-metering identifies unnecessary equipment running at night or during the weekends, provides the ability to provide usage evidence to the operators and facility managers the same day and to provide operators with feedback the next day about implementing changes. With sub-metering, all areas and equipment in the building that use large amounts of energy are monitored. Energy meters for all main electrical equipment per floor and for lighting and minor power use, such as plugs, are linked to the Building Management System (BMS) installed in case study B, which saves data and can find averages and trends, enabling users to know when there is a problem in the lighting or power systems of specific areas.

Automated Building Management System (BMS): This system is used in monitoring energy and water use and efficiency. The BMS is installed in case study B to regulate the air-conditioning and lighting system to prevent energy wastage.

4. Equipment specification

Electrical, lighting and mechanical equipment in buildings are usually specified based on design calculations by the mechanical and electrical engineer and also in compliance with standards, policies and regulations subsisting in the country in which the proposed infrastructure/building project is located. Many buildings particularly old ones still make use of electrical, lighting and mechanical systems that are not energy efficient and eco-friendly.

http://dx.doi.org/10.5772/66363

South Africa's electricity sector is regulated primarily by the National Energy Regulator of South Africa (NERSA), with the Department of Energy (DoE) as the custodian department. A number of standards and policies have been developed in South Africa to guide electrical equipment specification in buildings [4].

The key standards specified include the following:

SANS 10400-XA:2011 with SANS 204: These construction standards require that all new buildings and extensions to buildings are compliant with energy efficiency and energy use before receiving municipal approval.

SANS 941—Energy efficiency, energy performance and labelling of electrical and electronic apparatus: This standard covers energy efficiency requirements, measurement methods and appropriate labelling of energy efficiency electrical and electronic apparatus. This standard was published to ensure that at the time of purchase, buyers have all the relevant energy consumption information at their disposal. It also has implications for manufacturers and importers.

SANS 1544—Energy performance certificates for buildings: This standard specifies the methodology for calculating energy performance in existing buildings.

SANS 50010—Measurement and verification of energy savings: This standard was published in 2011 and specifies the methodology for calculating energy savings. This is a required tool for calculating savings for projects submitted on the 12L energy efficiency tax rebate programme.

VC9008—Compulsory specification for energy efficiency and labelling of electrical and electronic apparatus: This specification makes the SANS 941 a compulsory standard. It requires that a range of electrical and electronic apparatus adhere to certain minimum energy performance standards. It also requires that all appliances listed display the energy efficiency rating on the appliance.

Proposed smart metering standard—NRS 049 (advanced metering infrastructure): This specification will provide a standardised approach for municipalities and the electricity supply company of South Africa (Eskom) to follow. It will serve to describe the 'smart systems' which have been mandated for use by certain consumers in the Electricity Regulation Act.

The design team of case study B observed that lighting is the greatest energy consumer when compared to other building elements such as the air-conditioning system (as shown in **Figure 12**). Lighting is a continuous load that consumes a lot of energy. The design team therefore proposed a reduction of 94% in energy consumption in the case study B by introducing automatic lighting sensor systems an improvement on the requirements of SANS 204 of 2008, sub-metering systems and BMS.

The purpose of SANS 10400-XA:2011 with SANS 204 and SANS 941 is to ensure that buildings are more energy efficient, while SANS 1544, SANS 50010, VC9008 and NRS 049 is to ensure that energy consumption in buildings is effectively managed. Energy-efficient buildings translate into energy savings and less pressure on the public power supply and national grid, a reduction in power outages and rolling blackouts and ultimately the need for government to build new power stations, the attendant impact of power generation plants (either nuclear or coal powered through carbon emissions) on the environment is also reduced. The selected buildings complied with the key standards that guide electrical installation that aims at reducing energy consumption and conserving energy. For instance, case study A, B and C installed systems—wind turbines, regenerative lifts, geothermal loop systems and smart systems, with specifications above the minimum required is shown in **Figure 12.**

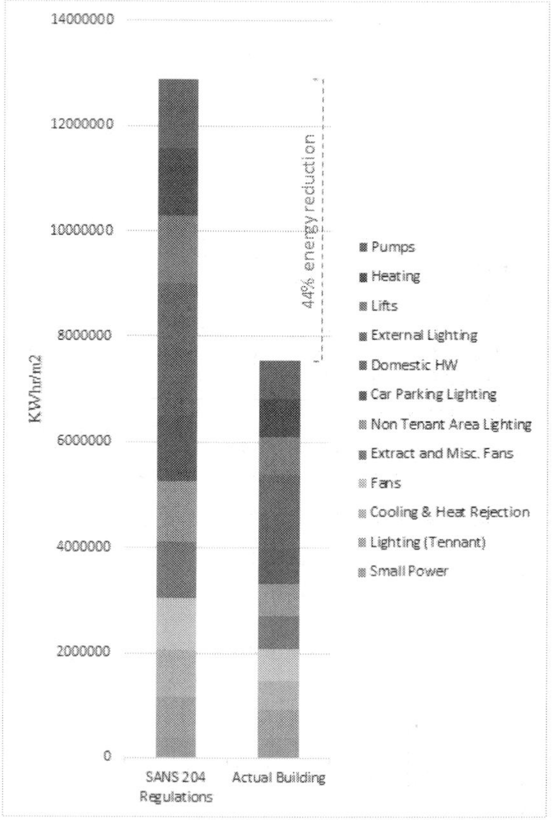

Figure 12. Energy consumption based on SANS 204 requirements vs. energy consumption in case study B.

Compliance with the standards outlined above will make it possible for buildings to attain targeted reduction in energy consumption when compared to conventional buildings that were

http://dx.doi.org/10.5772/66363

constructed before the South African National Standards (SANS), which specifies minimum requirements for energy efficiency in buildings. The selected buildings targeted significant reduction in energy consumption. While the project team for case study A targeted a 25% reduction in energy costs above conventional building standards, those of case study B and D targeted 56% and 30% reduction in energy costs, respectively. Implementing and going above the minimum standards saw a substantial change in energy consumption in the selected buildings.

The key policies on energy efficiency include:

Energy White Paper of 1998: This paper requires energy policies to consider energy efficiency and energy conservation within the framework of the Integrated Resource Plan (IRP), in meeting energy service needs from both the supply and demand side. It identifies the need for demand side management and the development and promotion of energy efficiency in South Africa.

NEES 2005, (2008): This policy sets out a national energy efficiency target of at least 12% by 2015. Sector targets range from 9% for transport, through to 15% for industry, commerce and the public sector.

Integrated Resource Plan (IRP) 2010: Specific targets for renewable energy and energy efficiency are set out in the IRP 2010 policy. This policy's revised balanced scenario provides an overview of the proposed new-build options including renewable options. The policy also outlines the energy savings expected from demand-side management programmes.

A key regulation includes:

The National Energy Act (Act 34 of 2008): This Act was legislated in support of economic growth and poverty alleviation in South Africa, by ensuring that diverse energy resources are available to the economy in sustainable quantities and at affordable prices. The Act is in pursuant of the development of the Integrated Energy Plan (IEP). It takes into account the interactions among economic sectors and environmental management requirements.

5. Power demand and energy consumption in buildings and its relationship to energy bills

Power rating of electrical appliances is usually measured in Watts (W) whilst the rated demand is measured in Kilo Watts (kW). The unit of electricity consumed is measured in Kilo Watts Hour (kWH).

Therefore, if the rated demand of an electrical appliance is known and the length of time for which it is used is also known, it is possible to calculate the cost of the energy supplied to the consumer in order to reduce expenses made on electricity. **Table 1** shows the power demand and energy consumption of some domestic appliances.

S/No	Appliance	Type	Rating	Rated demand in kW	Rated current in AMPS	No of hours for 1 kWH
1	Lamp		40 W	0.04	0.2	25
2			60 W	0.06	0.25	17
3			100 W	0.10	0.4	10
4	Iron	Small	750 W	0.75	0.75	1.3
5		Medium	850 W	0.85	3.6	1.3
6	Toaster	Regular	1000 W	1.0	4.3	1
7	Kettle	Small	2000 W	2.0	8.6	0.5
8		Medium	3500 W	3.5	11	0.3
9	Water heater	50 L	1200 W	1.2	5.2	0.8
10		100 L	2500 W	2.5	11	0.4
11	Cooker (four plate)	Regular	8000 W	8	34	0.12
12		Large	10,500 W	10.5	45	0.1
13	Single plate cooker	Portable	1800 W	1.8	7.7	0.6
14	Fan	Table	0.08 HP	0.06	0.25	17
15		Ceiling	0.3 HP	0.22	0.9	4.5
16	Air-conditioner	Small	1.5 HP	1.1	4.7	1
17		Medium	2 HP	1.5	5.6	0.7
18	Refrigerator	Small	0.2 HP	0.15	0.6	7
19		Medium	0.25 HP	0.19	0.8	5
20		Large	0.3 HP	0.22	0.9	4.5
21	Transistor radio		5 W	0.005	0.02	200
22	Stereo system		100 W	0.1	0.4	10
23	TV (black & white)		200 W	0.2	0.9	5
24	TV (colour)		300 W	0.3	1.3	3
25	Vacuum cleaner	Small	700 W	0.7	3.0	14
26		Medium	900 W	0.9	3.9	1
27	Water pump	Small	745 W	0.745	3.2	1.3
28	Washing machine	Automatic	600 W	0.6	2.5	2
29	Ditto + heater	Automatic	3000 W	3	13	0.3

Calculated from: Watts = Amps (current) × Voltage.
And 1 HP = 0.75 W.
Source: Windapo and Windapo [11].

Table 1. Power demand and energy consumption of some domestic appliances.

http://dx.doi.org/10.5772/66363

6. Determination of equipment that aid energy efficiency in buildings

Using the power ratings marked on the electrical, mechanical and lighting equipment (made mandatory as labels by SANS 941 in South Africa), it is possible to determine the equipment that optimises energy efficiency in buildings. The energy consumption level of the equipment can be compared to that of other comparable equipment. This can form the basis of value management and bringing about the improvement of energy performance in existing buildings.

7. Conclusion

This chapter examines the common equipment used in electrical, mechanical and lighting systems in four contemporary commercial buildings (offices and hotels) in Cape Town that aids energy efficiency. It was observed that photovoltaic panels, three 3 kW wind turbines and regenerative lifts were used in case study A as sources of energy generation, while geothermal loops and occupancy sensors were used in energy conservation. In addition, it was found that case study B employed an efficient lighting and sub-metering systems; case study C made use of a sea water cooling system, displacement ventilation, occupancy sensors and a BMS system; while case study D made use of LED lighting system and an efficient air-conditioning system. The study determined that the equipment used in these contemporary buildings are compliant with key specifications and policies guiding energy efficiency in buildings in South Africa. The chapter also presents an overview of power demand and energy consumption in buildings and an insight into how this is determined for household appliances and its implication for value management and energy efficiency initiatives. It was also observed that owners of the case study buildings were able to obtain significant reduction in energy consumption and running costs as a result of the installation of the equipment outlined above. Based on these findings, it can be concluded that energy-efficient equipment, which reduces energy demand in buildings, are widely adopted within contemporary commercial buildings in South Africa, in compliance with key standards, policies and regulations promulgated in South Africa to guide electrical equipment specification in buildings, with the main goal of achieving improved energy efficiency and reduction in energy consumed by buildings and by extension reduction in utility bills. This will provide significant savings for commercial buildings and private property owners.

Further studies that compares the energy demand of the equipment identified in this chapter to more conventional equipment is required. This will provide evidence to support the long-term energy savings/sustainability initiatives proposed by the government and private sector, and to present a business case for the use of these equipment in buildings.

Author details

Abimbola Olukemi Windapo

Address all correspondence to: abimbola.windapo@uct.ac.za

University of Cape Town, Rondebosch, South Africa

References

[1] GreenCape. Energy Services: Market Intelligence Report 2016. Cape Town: GreenCape; 2016.

[2] Medium-Term Forecasting Associates (MFA). Report on Building Costs. Medium Term Forecasting Associates. Stellenbosch: MFA; 2015.

[3] Langdon and Seah. Construction Cost Handbook [Internet]. 2015. Available from: www.langdonseah.com/en/ga/files/download/3616 [Accessed: 2016-08-06].

[4] Mobley P. Market Analysis: 2010 Experience Exchange Report indicates Tight Management Needed for Today's Demanding Market. The BOMA. Kinsley REPORT: BOMA International [Internet]. 2010. Available from: www.boma.org/research/newsroom/boma-kingsley/Documents/BKR_Autumn2010.pdf [Accessed: 2016].

[5] Hodge N. Power Trip, Energy Risks, Special Topic on Emerging Risks [Internet]. Undated. Available from: www.agcs.allianz.com/insights [Accessed: 2016-07-18].

[6] Kufeoglu S, Lehtonen M. Interruption Costs of Service Sector Electricity Customers, A Hybrid Approach. International Journal of Electrical Power and Energy Systems. 2015; **64**:588–595. DOI:10.1080/014461998372231

[7] Heyns F. The Sky's the Limit. *Earth Works Magazine* [Internet]. 2013. Available from: http://greencape.co.za/assets/blogcontent/201305/Allan-Gray-small.pdf [Accessed 2014-09-18].

[8] Makai Ocean Engineering. Cold Seawater Air Conditioning [Internet]. 2014. Available from: http://www.makai.com/pipelines/ac-pipelines/[Accessed 2014-09-18].

[9] Cooke J. Architecture South Africa. Journal of the South African Institute of Architects. 2015; July/August Issue:1–62.

[10] Turner Group. Displacement Ventilation Design [Online]. 2014. Available from: http://www.turnerbuildingscience.com/hvac-displacement-ventilation.php [Accessed 2014-09-19].

[11] Windapo A, Windapo B. Building Services Study Guide. Cape Town: University of Cape Town Libraries; 2014. pp. 83–84.

Chapter 9

Towards Near Zero Energy Home

Esam Elsarrag and Yousef Alhorr

Additional information is available at the end of the chapter

http://dx.doi.org/10.5772/66224

Abstract

In the context of building design, as investment in the built-environment continues to grow, the requirement to deliver low-energy buildings will become ever more pressing as natural resources dwindle and consumer energy costs increase. Energy efficiency awareness and regulations continue to rise in the Gulf Cooperation Council (GCC) countries but the majority of building stock of which the larger share in energy consumption has not been designed for energy efficiency. The design and construction of buildings in hot humid climates require high-energy consumption typically for air conditioning due to higher thermal loads. Regionally, there is a rising concern on the current rate of energy consumption due to air conditioning. The global sustainability assessment system (GSAS), a performance-based system raised the bar of efficient design by the development of stringent energy passive design benchmarks on the thermal cooling need in buildings. This chapter introduces the simulation measures undertaken to reduce the cooling need using a 'showcase' house or the 'near Zero Energy Home' (nZEH), which is currently under construction. The chapter presents and discusses the Be Lean, Be Clean and Be Green strategies that used to reduce the cooling demand by more than 80% and the overall energy consumption by 75%.

Keywords: energy efficiency, zero energy home, renewables, solar cooling, climate change

1. Introduction

As the population continues to grow, there will be increased pressure and demand for energy resources. Considering the wider impacts of carbon emissions on our climate, effective energy efficiency solutions are necessary in order to achieve the overall goal of reducing these emissions.

Qatar and in particular the Gulf Cooperation Council (GCC) countries, in general, can be characterised by an extreme set of climatic conditions regarding the challenges under consideration. These conditions relate, among others, to aspects identified in the literature [1–3]. Extreme climatic conditions impose a heavy reliance on cooling, mostly electricity-based, and thus a strong and structural dependency of a high-energy resource. Hourly outdoor dry temperature variations during a year in Doha, Qatar are shown in **Figure 1**. The average of highest outdoor temperatures during a year is 37.03°C, however, high-temperature values that exceed 46°C could be observed in summer. As shown in **Figure 2**, the temperature exceeds 40°C for more than 300 hours, which is anticipated to be doubled when considering Doha climate change, by 2025.

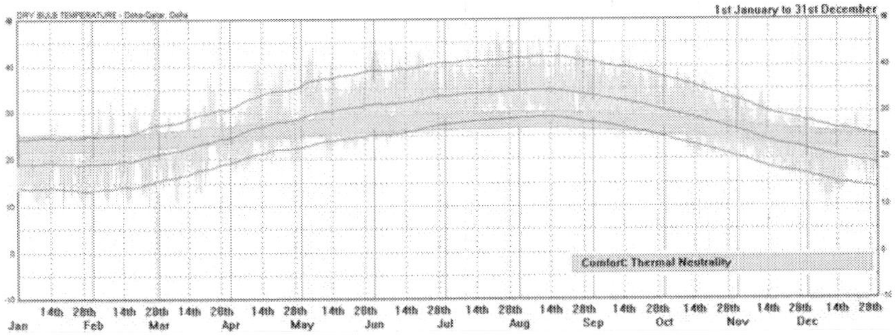

Figure 1. Outdoor dry bulb temperature variations.

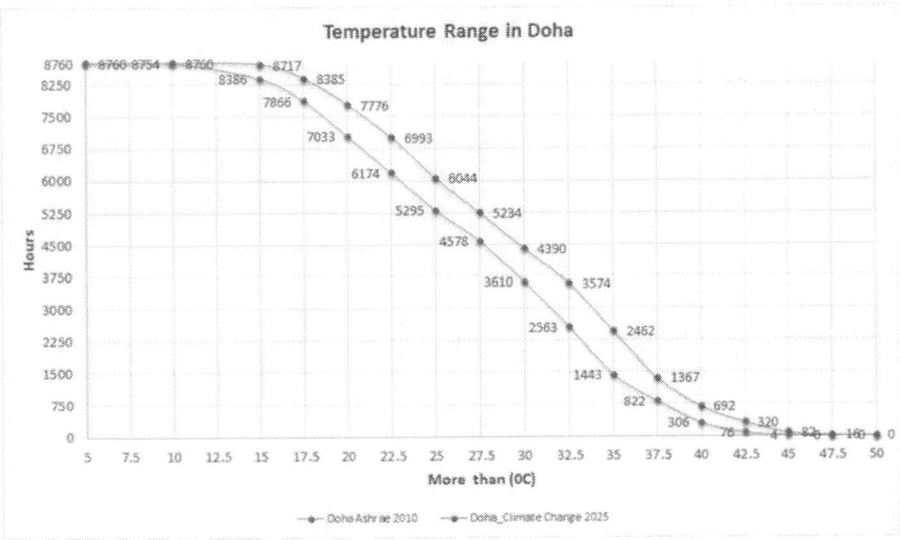

Figure 2. Temperature range in Doha.

http://dx.doi.org/10.5772/66224

The hourly outdoor air relative humidity (RH) variation is shown in **Figure 3**. The data show that the daily average RH is always greater than 40%, and the daily maximum RH often goes above 80%. Dehumidification by air-conditioning is required all year round.

Figure 4 shows solar radiation by different orientations. Annual cumulated solar radiation amounts from different orientations are different. The data show that the west and south walls of a building receive the most solar radiation, less than the roof with an annual average of 54%.

Many buildings in the GCC have been constructed following international models that are not fit for the particular conditions imposed by the local context. For instance, the design of modern high-rise buildings, with unfortunate high glazing to wall ratio, increased dramatically the energy consumption due to high solar gains. In addition to insufficiently insulated building skins, a lack of passive design measures for energy consumption control such as glazing and shading features and an excessive proneness to overheating result in an excessive

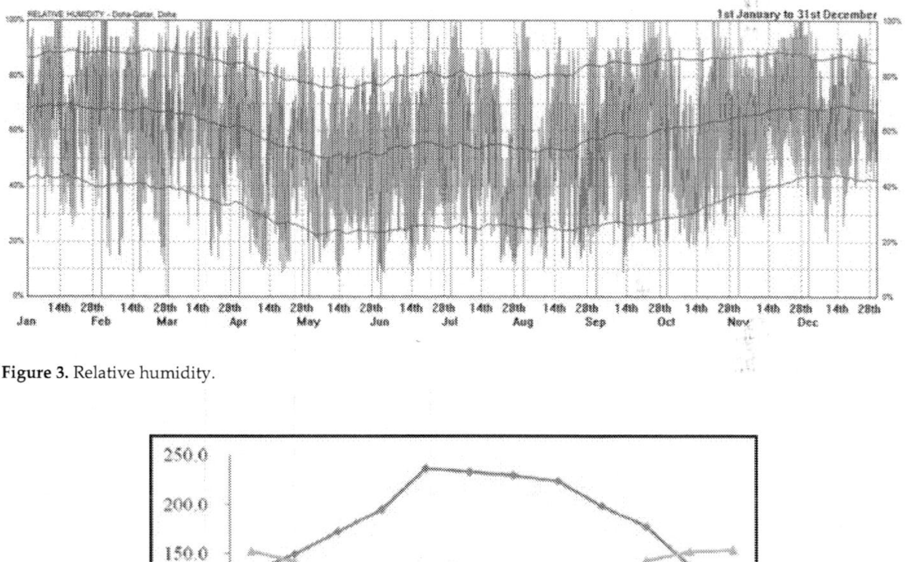

Figure 3. Relative humidity.

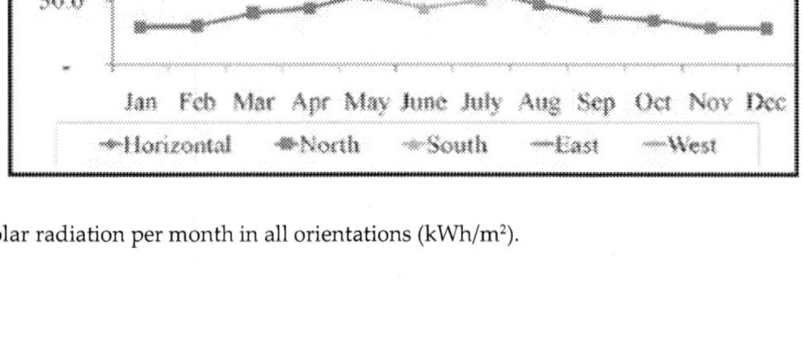

Figure 4. Global solar radiation per month in all orientations (kWh/m²).

reliance on active indoor climate control. Air conditioning counts for more than 60% of the electricity consumption in the GCC countries [4]. Moreover, this lack of responsiveness to the local climatic conditions also leads to problems of indoor air quality, user comfort and user productivity.

With energy being cheaply available, the incentive for building users to save on their energy consumption is weak. This is moreover the case for three major contributing factors at the same time: building skins, HVAC installations and user behaviour. As a consequence, extremely high-energy consumption and greenhouse gas emission figures. According to the IEA records, Qatar has both the highest CO_2 emissions and energy consumption per capita in the world [3].

In Qatar and the GCC, not only the total annual energy consumption in buildings is very high, but peak demands for electricity also put a heavy burden on the infrastructures needed to respond to such demand pattern. In this context, building energy efficiency strategies can help to realise peak shaving by load reduction and load shifting.

As stated earlier, rapid population growth increases the stress on the energy infrastructure system, and is particularly related to the peak load and installed capacity problem.

Thermal regulation of the indoor spaces can be accomplished by simple passive design methods to utilise the natural sources of energy, such as the sun and the wind to provide domestic hot water (DHW), cooling, ventilation and lighting and to contribute to responsible energy use. During recent years, building construction continued to bloom in the Gulf region. However, the hot and humid climate of the region makes homes demand large quantity of energy for air-conditioning to achieve an acceple thermal comfort level. On a yearly basis, the cooling load dominates the building thermal load, while the heating load can be negligible for the energy use estimation in buildings. The weather in Qatar is moderate only during 3 months, while in most of the other months, it is warm/hot or even very hot and humid [1].

The concentration of carbon dioxide and other greenhouse gases in our atmosphere is now recognised as the driver of global warming and climate change. The Intergovernmental Panel on Climate Change (IPCC) advised that cut of 25–40% by 2020 would be necessary, compared to a 1990 baseline, to limit the global atmospheric average temperature rise to 2°C. Despite these publications, global emissions have increased to 40% above the 1990 baseline by 2009 and the concentration of atmospheric CO_2 is now believed to be at its highest for at least 800,000 years [5].

Energy benchmarks were introduced for several new building types in Qatar by the Global Sustainability Assessment System (GSAS) [6], as given in **Figure 5**.

The worldwide drive towards curtailing carbon emissions and improving the sustainability of our social, economic and cultural networks is now well underway. Considering that cooling in buildings accounts for two-thirds of all energy consumption in buildings, the use of passive design to limit the need for cooling, and thus the electrical consumption associated with cooling, is likely to lead to a significant overall reduction in carbon emissions.

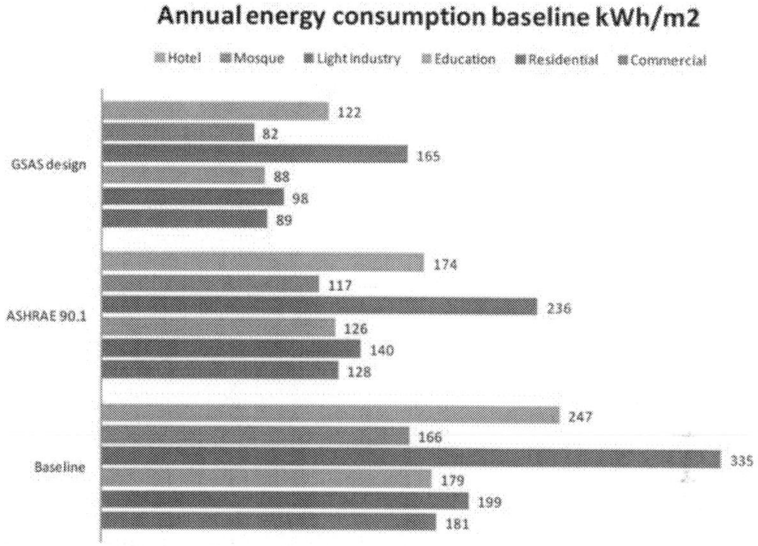

Figure 5. Energy consumption baseline for different new build in Qatar (kWh/m^2/year) [6].

There has been a very little work available about the design of zero energy home in hot/tropical regions [7, 8]. The literature lacks in providing similar examples from the MENA region and hot arid climates. Sustainability venture presents obstacles and it appears to be challenging, especially to find data on several countries [9]. Therefore, this chapter is coming to bridge the gap through a comprehensive study of dynamic design. Simulation studies and analysis, calculations and a real demonstrative building are developed. In particular, the near zero energy home (nZEH) project will try to focus on efficient design solutions for residential buildings as well as the design considering the influence of the local climate, the endogenous energy resources and the local economic conditions. Emphasis will be given to the bioclimatic design of the house and the renewable energies utilisation in order to achieve objective towards nZEH.

2. Background

Conventionally, the main energy generation methods are by the combustion of fossil fuels which are limited resources on the Earth. Excessive burning of fossil fuels also intensifies the global greenhouse gas content, which accelerates the global warming effect. As a result, a novel building-design concept namely, net or near net zero energy building (NZEB) has emerged, which aims to reduce the energy demands of buildings from fossil fuels resources and to magnify the utilisation of renewable energy in buildings.

In general, NZEB is defined as a building with zero net energy consumption, which means that the total amount of energy consumed by the building is roughly equal to the amount of renewable energy generated on the building site on a year-round basis [10, 11]. More definitions of NZEB based on different evaluation principles and across different countries were presented [12]. A strategic goal of the building technologies program of the US Department (DOE) is to accomplish 'marketable zero energy homes in 2020 and commercial zero energy buildings in 2025' [13]. In Europe, in the recast of the EU Directive on energy performance of building (EPBD), it was specified that all new buildings shall be 'nearly zero energy building' by the end of 2020 [14]. *Furthermore, since October 2008, a panel of experts from International Energy Agency (IEA) has initiated a project entitled 'Towards Net Zero Energy Solar Buildings' [15] which was intended to analyse exemplary buildings that were near a zero-energy balance in order to develop methods and tools for the planning, design and operation of such buildings.*

More efforts should be put on the research of near zero energy home (nZEH) in hot climate geographical location, especially in the Gulf region as it *is lacking* comprehensive research in near zero energy buildings.

Unfortunately, the lack of a common approach for this new type of buildings results in misunderstandings, endless discussions and different solutions per project [16]. Recent studies note that; despite the *exciting* phrase of 'zero *energy*', near zero energy building definition often lacks a clear and widely accepted explanation of what this term *actually* means. Researchers indicate that the definition of zero energy buildings concept can be constructed in several ways, depending on the project goals, intentions of the sponsors, concern about the climate changes and greenhouse gas emissions and finally, the energy costs [11, 17]. Smart technologies, efficient structure *design* and innovative materials should follow a robust legislation contributing to healthier and more *efficient* living.

Improving the energy efficiency measures has taken the attention of researchers in *hot*, humid climate in the Gulf region. Experimental and theoretical studies were conducted recently to improve building fabric efficiency and promote enhanced indoor air quality in *hot*, humid climates [18–20]. Energy and carbon framework model has been developed and implemented in the Gulf region [1], as shown in **Figure 6**. Passive design of high-rise buildings attracted researchers from different parts of the world [21–25].

Figure 6. Proposed energy and carbon model structure [1].

In China, passive design zones for different climates in18 cities representing the 5 major climate types *were developed*. A bioclimatic approach was adopted in which the comfort zone and 12 monthly climatic lines were determined and plotted on the psychrometric chart for each city. The potential use of nine passive design strategies *was assessed* and passive design strategy zones were identified [26].

Another study presented an integrated approach to explore how energy consumption could *be minimised* in residential buildings by optimising seven passive design measures for 25 representative cities. The study showed that, with *optimisation*, the passive design could reduce *annual* thermal load of building considerably and could replace air-condition systems in winter for the areas with high solar radiation [27]. Within a system transition, for nZEH, a complex interplay between technical and non-technical challenges emerges. Consequently, a holistic approach is fundamental for success when it comes to practical implementation.

Technical challenges concern the best trade-off between available passive design solutions, the development of new concepts, the integration with technical installations and energy and mobility systems at different scale levels (building, neighbourhood, region or state), and the interplay with indoor environmental requirements (air quality, thermal comfort, health and productivity). Relying more on passive design measures and reducing the share of active climate control is another technical challenge, which however also has a strong cultural component. Financial aspects may be considered as another set of technical parameters. Life cycle cost or net present value will be important aspects to consider, both for selecting the economically most promising options and for providing decisive arguments to the potential investor. The identification of feasible business models is crucial for any development to become successful in reality.

Non-technical challenges concern, for example, supporting energy efficiency uptake by policy measures such as regulation or subsidising, training building sector actors to assimilate new techniques and passive design concepts, and educating the public at large about the benefits of energy efficiency and building renovation.

Socio-cultural aspects are another important, non-technical matter of concern. As nZEH, in particular in the residential sector, intervenes deep into the life of citizens, it shall be sensible to this two-sided challenge. Reconnecting with the environmental wisdom and cultural roots of traditional architecture may deliver unique concepts in the Qatari or the GCC context. We can see that this tendency is emerging in new architecture (such as the Burj Doha tower, the Msheireb and Lusail developments in Doha or the Al Bahar Towers in Abu Dhabi).

Researchers in the Gulf region [1, 4, 6] developed an energy hierarchy and integrated multi-level approach that connects sustainable developments from the national level to the project level to reduce the carbon emissions in Qatar, see **Figure 7**. The implementation of the strategy in a city in Qatar forecasted more than 30% reduction in CO_2 emissions compared to the existing standards, see **Figure 8**.

Previous research [28] showed that the average annual solar radiation falling on the Arabian Peninsula was about 2200 kWh/m^2. Other researchers [29] also concluded that Saudi Arabia, a

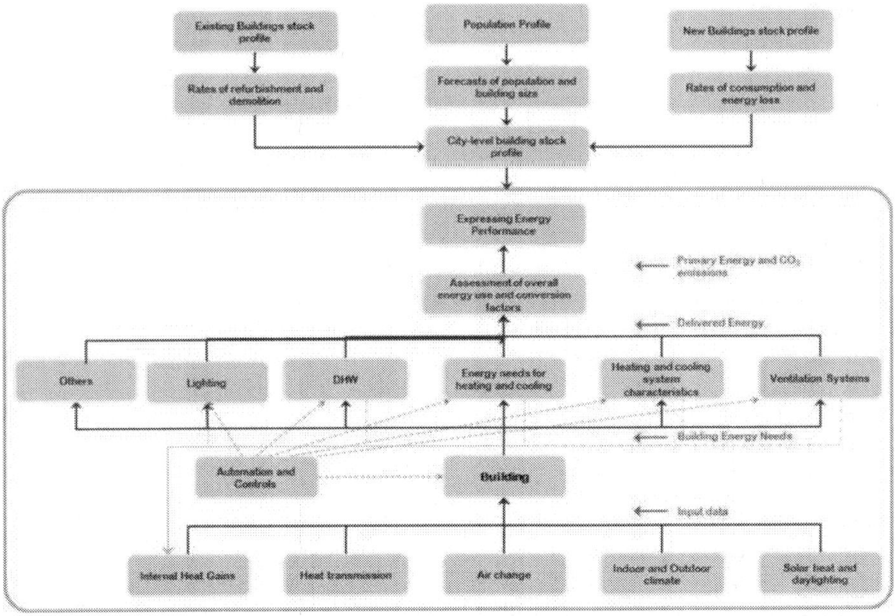

Figure 7. Passive climate design measures have always been a fundamental asset of traditional architecture.

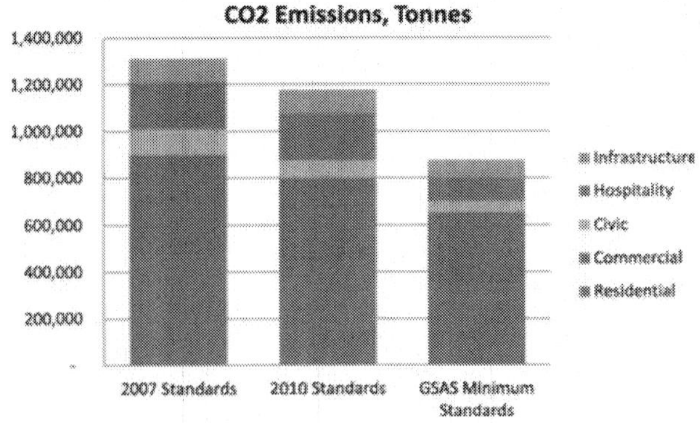

Figure 8. Forecast impact of GSAS standard on CO_2 emissions from new cities in Qatar by 2020 [1].

country located in the Gulf region and adjacent to Qatar has great natural potential for solar power generation and economic incentives to develop renewable energy to meet domestic electricity demand. Both studies further consolidate the fact that solar energy is abundant in Qatar and the Gulf region. This implies that under proper design, the energy generated from solar energy is possible to cover part to full demand in a residential unit.

An experiment of passive-house with dynamic insulation was performed in the Gulf region [4, 18, 19]. The experiment results revealed that dynamic insulation could reduce 17% of cooling energy demand as compared with a static insulation. The findings are adequately encouraging to continue the research and development activities of dynamic insulation technologies, which can be applied into the nZEH.

Apart from the dynamic facade, a novel titled solar liquid desiccant regeneration system, which potentially reduces the cooling demand of an indoor space, was also studied [30, 31]. The study provides a good reference for such system to be employed in the nZEH. Simulation studies on the performance of solar absorption cooling system in another country in the Gulf region (UAE) were performed [32]. The simulation results showed that the solar absorption cooling system was operated at optimum COP and the chilling capacity was sufficient throughout the year except June–September. The insufficient cooling capacity over the June–September period could be improved by increasing the gross area of the solar collectors. Therefore, solar absorption cooling system may also be a desirable option to be utilised in the nZEH. Detailed literature reviews and descriptions of the evaluation methods for net zero energy building can also be found [33].

In the current state of affairs indeed, buildings are more and more becoming smart entities operating in smart networks, in particular when it comes to renewable energy provision and the deployment of smart grid infrastructures. Buildings hereby become an agent in the energy infrastructure, consuming, producing or buffering different forms of energy in order to arrive at optimal functioning both at the building level and the district level. Accommodation of intermittent renewable energy production, smart control algorithms, demand side management and buffer capacity are some of the main pillars of this new energy paradigm. A good example for the Qatari context would be the deployment of fourth generation smart cooling grids based on renewables like the Sun or geothermal energy, in combination with adiabatic cooling conversion.

It is important to note that recent experiences in the GCC indicate that in a smart and renewable energy infrastructure paradigm, energy efficiency turns out to be a critical factor for the feasibility of the former [34]. Energy demands and peak loads tend to be so high that, in order to fulfil all demands with renewable sources, energy efficiency remains paramount. Unlike the NZEB approach of temperate climates, extra energy consumption for indoor space cooling is expected for the nZEH in the Gulf region.

3. The nZEH description and energy hierarchy

To this end, the aim is to present an exceptional prototype building that may support the future building sector towards nearly zero energy solutions. The proposed home design concept will be a progression form of passive and bioclimatic sustainable design. A performance measure for the nZEH will be implemented during the design process and a real-time performance indicator will be derived. The operation process after the completion of the construction will be measured and monitored using a smart metering solution. This dynamic

Figure 9. The eco home perspective.

measurement can be integrated on seasonal or yearly basis to show the overall performance. The eco-house will be a landmark project that will reinforce Qatar's reputation as a leading advocacy of sustainable development and will drive in a new generation of nZEH building in the region. This house will offer a genuinely sustainable, smart and healthy living environment for residents. A number of methodologies will be carried out through this 5-year project duration in order to achieve the aim of the nZEH.

Public and private sector of Qatar already have started the implementation of strategies towards minimising the carbon footprint of the building sector and improving the quality of life. GSAS, a performance-based sustainability rating system, is developed by the Gulf Organisation for Research and Development [35]. GSAS is studying the local situation in the Gulf region, the weather, the local standards and practices and many more, leading to the formulation of value statements and will help the project throughout the whole design and construction process to include the most efficient strategies and techniques [36]. GSAS limited the maximum annual cooling demand for new-build housing compliance in Qatar to 125 kWh/m^2 however, conventional designs exceed 250 kWh/m^2. The project will develop a comprehensive design and market search in this direction, which will lead in a financially acceptable, feasible and totally successful project. **Figure 9** shows that the architectural language reflected in the villa designed to award a modern interpretation of the timeless traditional Qatari architecture, reflecting the culture and heritage of Qatar.

The nZEH, a detached family home, was designed in the hot-humid climate in Qatar as a 'showcase' study using the following strategies: Be Lean; Be Clean and Be Green as shown in **Figure 10**.

3.1. Be Lean strategy: reduce energy need

The envelope has an impact on cooling load and day lighting considerations. It was modelled as a climate modifier rather than solely a means of excluding external climatic conditions. It consists of structural materials and finishes that make up the exterior of the building and separate the inside from the outside. The envelope will have the ability to minimise solar heat gain and avoid overheating, also to use window shading and thermal mass to attenuate heat

http://dx.doi.org/10.5772/66224

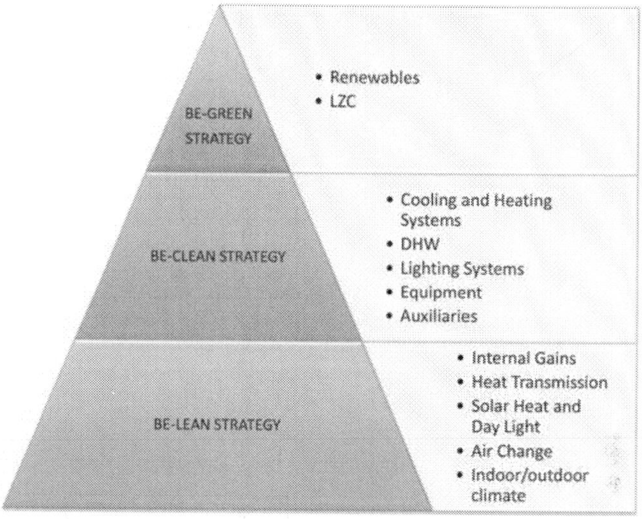

Figure 10. The nZEH strategies used in the nZEH showcase.

gain; to allow optimum levels of natural ventilation and day-lighting. It is worth noting that higher humidity ratios and dust are the main barrier of using natural ventilation, hence it is essential to use mechanical or a hybrid ventilation system in several cities within the Gulf region. A novel façade design approach has been modelled and will be implemented in the nZEH. It will have a compact design to reduce the influence of the external environment and may also benefit by need of less space for the distribution of horizontal and vertical services, particularly for air ductwork. Assemblies of air permeable adaptive insulation cells will be fitted internally over the available wall area and sealed in place using an independent, self-supporting gypsum board lining system that contain and protect the cells. As the name suggests, adaptive insulation permits the flow of air through the cells to facilitate the recovery of building fabric heat or, as in the present case, cool loss to ventilation air. Dynamic insulation uses the fabric as a heat exchanger. It either captures the heat loss from the envelope via the ventilation air in cold climates or rejecting the heat gain via the exhaust air in hot climates. Under steady state conditions, the heat loss through a conventional 'static' wall is the same entering as leaving the wall, its magnitude determined by the resistance to heat flow of the materials used to build the wall. By contrast, the parietodynamic wall in **Figure 11a** shows cold outdoor air being drawn in through the void created by the insulation layer and pre-heated by fabric heat loss, recovering part of that heat loss to the building as tempered air. Use of the wall to exhaust chilled indoor air provides an effective means of rejecting wall heat gain in the hot climate, to deliver precisely the same reduction in wall *U*-value as a function of airflow rate.

Researchers [37, 38] reported the results from the first field trial of dynamic insulation in Abu Dhabi. They also investigated the use of dynamic insulation in a building facade for local zonal insulation and ventilation [38].

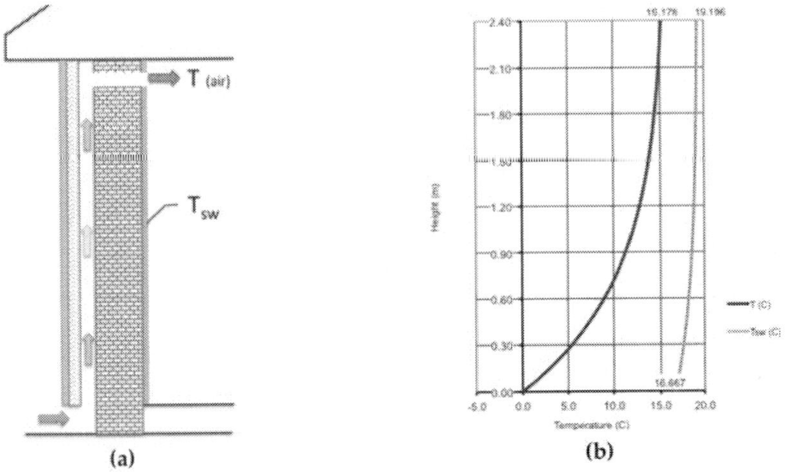

Figure 11. Solid wall system using 50 mm eps sheet (a) Sectional schematic. (b) Air and surface temperatures.

The height-averaged dynamic *R*-value of a dynamic wall or building façade can be estimated using the following steady-state expression [39]:

$$R_d = \frac{1}{U_d} = \frac{(T_i - T_o) \times N R_o}{(M - T_o)(e^{-N} + N - 1)}$$

where

$$M = \frac{R_o T_i + R_i T_o}{R_o + R_i}$$

$$N = \frac{R_o + R_i}{Rho_a C_a V_u R_i R_o}$$

T_i and T_o are indoor and outdoor temperatures, R_o and R_i are the aggregated thermal resistances (*R*-values) between the void space and the cladding to ambient and indoor interfaces respectively, Rho_a the air density, C_a the specific heat capacity of air and V_u the volume flow rate of air per unit width of wall. Equation ignores radiation, convection, thermal inertia and secondary void space effects.

A study for a dynamically insulated villa showed that the dynamic *U*-value is 0.125 kWh/m²K while the static is 0.24 kWh/m²K based on the conditions shown in **Figure 12** [40].

3.2. Be Clean strategy: reduce energy consumption

The use of efficient systems and effective means of control is vital to reduce the energy consumption. A study showed that the energy saving of a building heating system by adopting model predictive control (MPC) could reach the range of 15–28% [41]. At times when ventilation

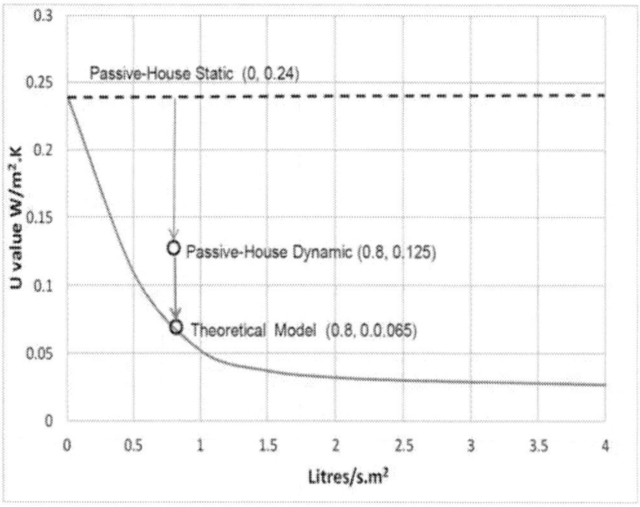

Figure 12. Dynamic *U*-values (W/m²K) [40].

and daylight cannot alone meet the needs of occupants, the building services should meet the remaining demands as simply and effectively as practicable, in harmony with the occupants and the building as a whole. An essential part of the integrated design is to ensure that the energy supply and monitoring strategy is as coherent and environmentally sustainable as possible.

Figure 13 shows about 83% of the ambient weather conditions are not in the comfort zone, therefore, the following strategies are used to design a high efficient cost-effective system.

The innovative efficient system design includes:

- Variable frequency drive, two-stage indirect evaporative air conditioner coupled with liquid desiccant dehumidifier and the energy recovery system
- Low-grade heat driven 17.6 kW absorption chiller (input temperature range of 75–85°C), coupled with 35.2 kW variable refrigerant flow (VRF).
- Controls play the vital role for the sequence and operation of the system especially when coupled with the dynamic façade and the HVAC system as stated above.

3.3. Be Green strategy: renewables

There have been numbers of high-energy efficiency system installations coupled with the building to potentially generate and store useful energy from renewable energy sources for the occupants. The 17.6 kW 'All in One' absorption chiller is fully driven by renewables. The solar absorption chiller is fully integrated with a hybrid cooler and a thermal store. Flat plate solar collectors, 50 m^2, are used to charge the thermal store (75–85°C) in addition to 4 kWp photovoltaic for lighting auxiliaries and small power. The nZEH includes smart meters and sub-metering to ensure that future building performance can be continually monitored by the building operator.

The preliminary data that provide a draft description of the nZEH are presented in **Table 1**.

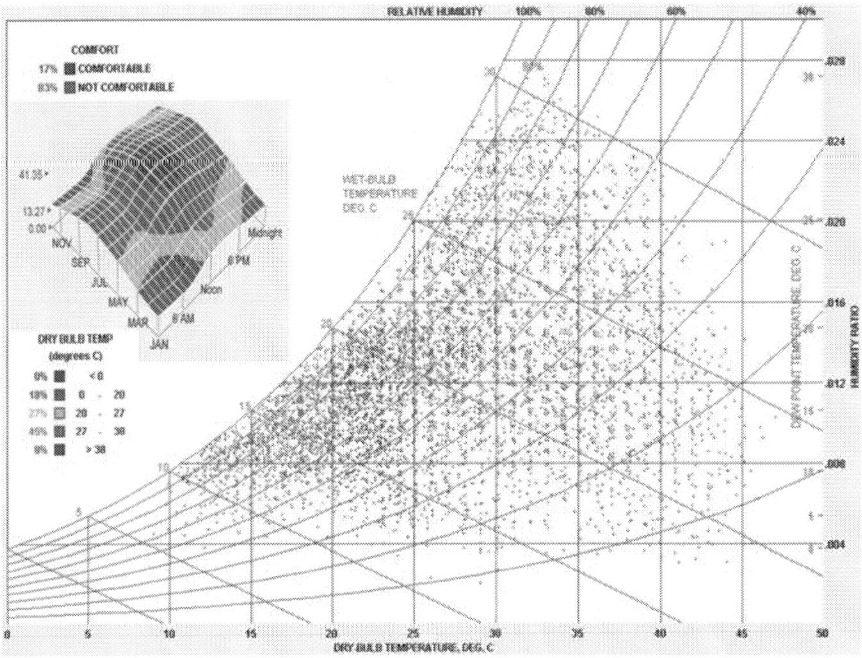

Figure 13. Extreme weather conditions of Doha city.

Gross area (m^2)		**733 m^2**
Number of bedrooms/occupants		7 persons
Space cooling design		VRF (35.2 kW), solar absorption chiller (17.6 kW)
Water heating design		Solar water heating, 50 m^2
Renewable energy		PV panels, 4 kWp
Lighting features		LED, 4.5 W/m^2
Ventilation		Controlled demand, maximum 400 L/s
Building Envelope Characteristics	**Roof**	*U*-values: 0.1 [W/(m^2k)]
	Walls	*U*-values: 0.24 static, 0.07 dynamic [W/(m^2k)]
	Windows	*U*-values: 1.3 [W/(m^2k)], SC = 0.25, external shading
Other Features		Socio-cultural, energy monitoring and control
GSAS rating target		6 stars

Table 1. Home specifications.

http://dx.doi.org/10.5772/66224

4. Results and discussion

All year round energy performance of the nZEH can be estimated by integrating the component modules of the associated system with the architectural design of the nZEH in the simulation platform. The design parameters' values of the associated systems and of the residential unit, as well as the control strategies which result in an energy balance of energy generation and energy consumption throughout a year, can also be obtained via the simulation exercises.

Figure 14 summarises the strategies undertaken to reduce the building's cooling need. The baseline is based on ASHRAE 90.1 and 90.2 (2010) and current local authority's regulations. Implementing conventional materials to the proposed design, the annual cooling need is about 180 MWh (246 kWh/m^2). This figure is in agreement with the literature (250 kWh/m^2). Applying the dynamic façade strategy will reduce the cooling need by 6.9%. Solar radiation is very critical, optimising the window to wall ratio, orientation, providing proper shading and the use of high-performance glazing can reduce the cooling need by 20%. The super-efficient design can reduce the cooling need by almost 50%. **Figures 15** and **16** compare the baseline and super-efficient hourly and monthly cooling demand. The passive measures reduced the peak dramatically and the loads are almost flattened.

Further to the passive design, efficient cooling systems and proper controls can be applied to reduce the need for electricity. As shown above, the passive measures reduced the cooling need by almost 50%. The most used air conditioning system for single residential homes in Qatar and the GCC countries is the split direct expansion unit with a typical coefficient

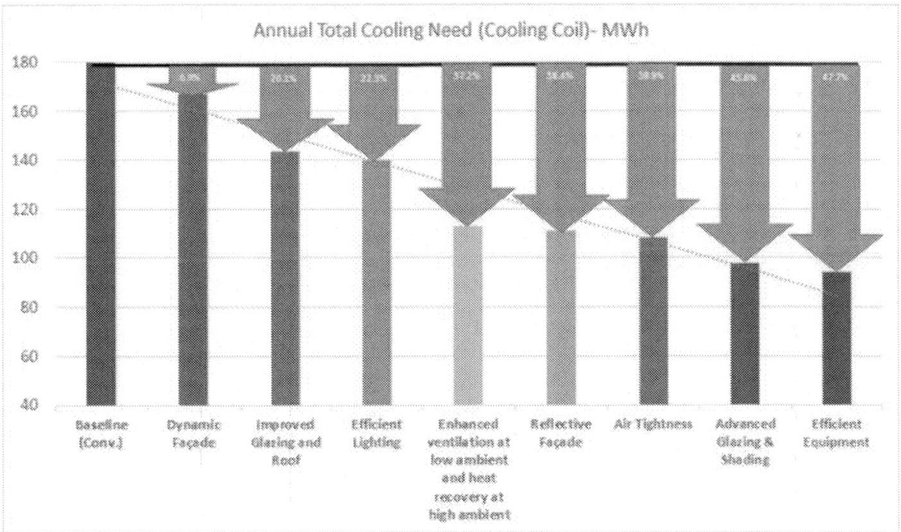

Figure 14. Annual total cooling need (cooling coil) MWh.

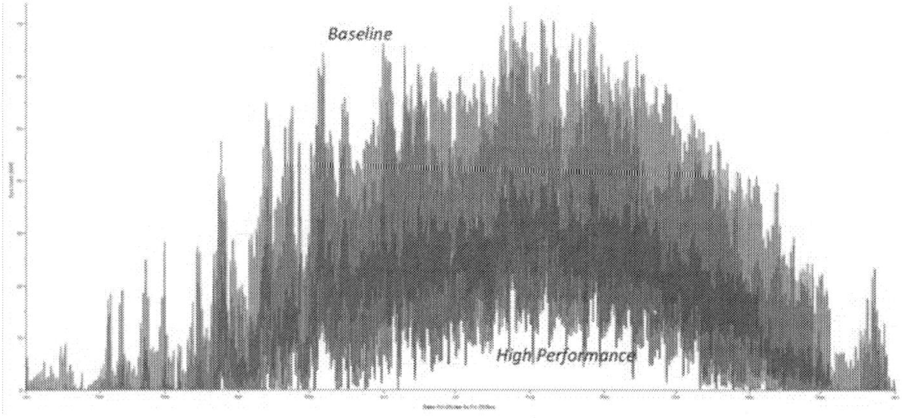

Figure 15. Hourly cooling loads (kW).

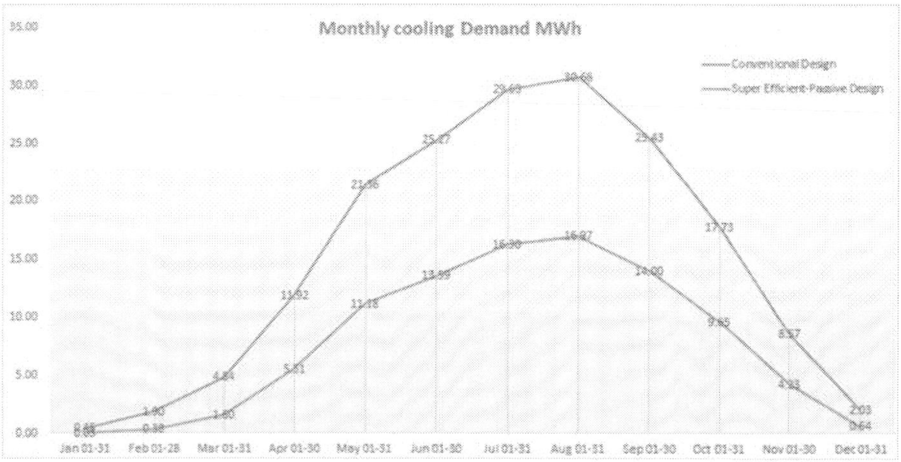

Figure 16. Monthly cooling demand (MWh).

of performance (COP) of 2. Moving to more efficient systems such as package systems or variable refrigerant flow will result in higher COP and hence reduce further the electrical demand. As shown in **Figure 17**, the use of VRF reduced the electrical need by 12%. It is anticipated that the energy performance of the nZEH can be further improved by the advancement of the control system. Model predictive control (MPC) with weather prediction is a high potential option to be applied in the continuation study. In the nZEH, indirect-direct

http://dx.doi.org/10.5772/66224

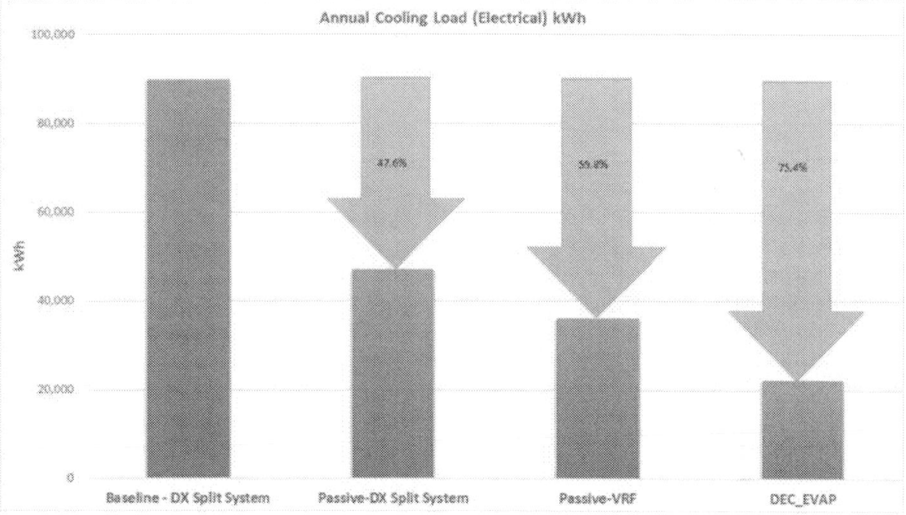

Figure 17. Annual cooling load (electrical) kWh.

evaporative cooling coupled with a desiccant system (DEC_EVAP) will be used to cool the house during certain periods. Also, it will be used to dehumidify the air when the humidity is high. Coupling such system will reduce the cooling loads by further 15% without compromising the thermal comfort. Such savings could be much higher if a new comfort zone is defined.

As shown in **Figure 18**, the use of low-grade heat 5TR (17.6 kW) solar absorption chiller (SAC) driven by 50 m^2 of solar water heater will reduce the cooling load by 8.5%. The integration of SAC with the domestic hot water demand will also reduce the needs of electricity to produce hot water, see **Figure 19**. Initially, the use of water efficiency measure (low flow fixtures) reduced water demand by more than 60%. As shown in **Figure 19**, the DHW is fully driven by the solar system.

Further for a clean strategy, renewable energy systems have to be applied in order to generate the remaining energy.

As shown in **Figures 20** and **21**, the use of 4 kWp to drive the auxiliaries and other small power is feasible. This will contribute an additional 3% reduction in annual energy consumption and an overall reduction of 75% compared to the baseline. Consequently, the house can be considered as near zero energy home (nZEH).

On the other hand, offsetting the 25% by renewables is not realistic and needs a lot of considerations. The house requires an extra 25 kW from renewables in order to achieve the net zero energy building, NZEB. In such case, the use of photovoltaic panels will require more than 200 m^2 of land or roof area which is not feasible.

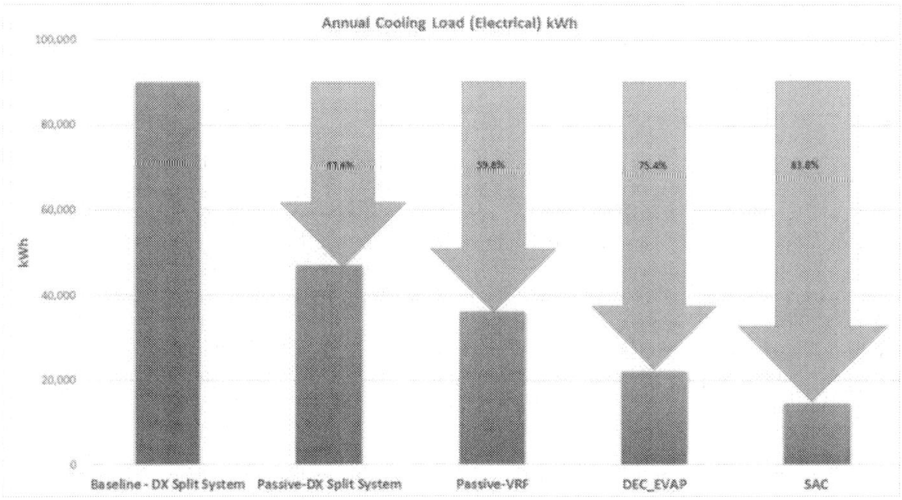

Figure 18. Annual cooling load (electrical) kWh.

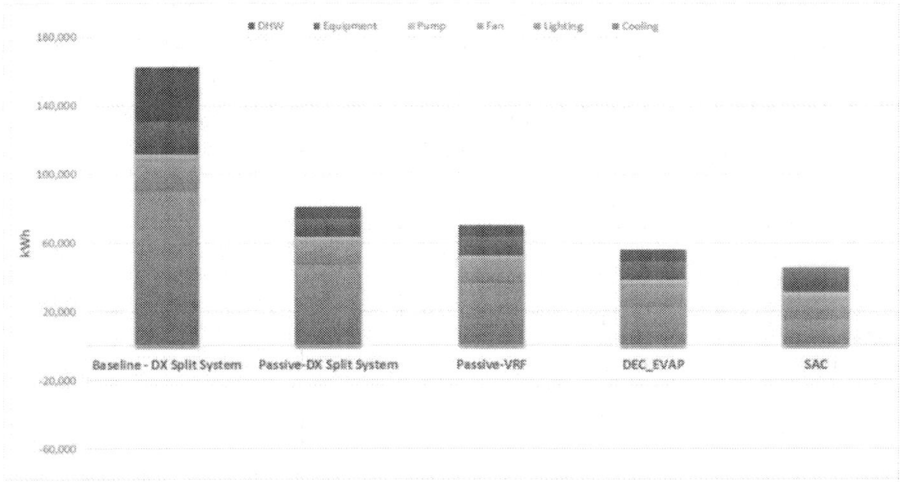

Figure 19. Total annual energy consumption by type using different strategies (kWh).

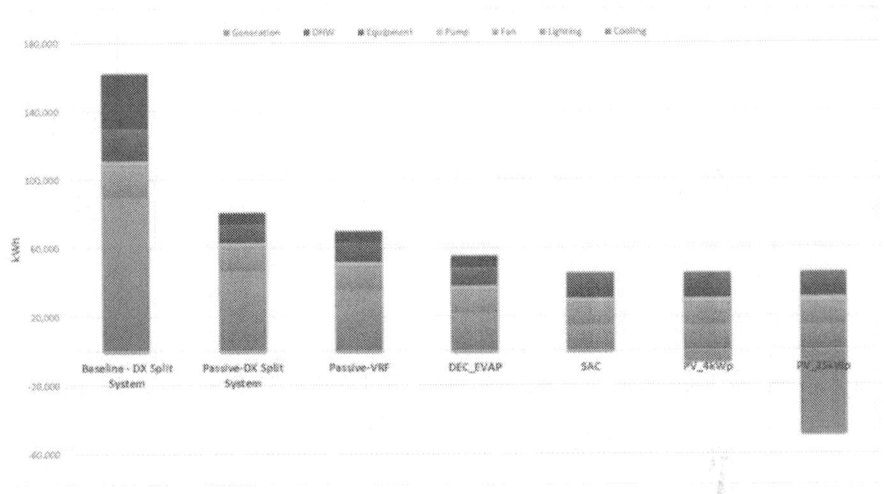

Figure 20. Total annual energy consumption by type using different strategies with renewables (kWh).

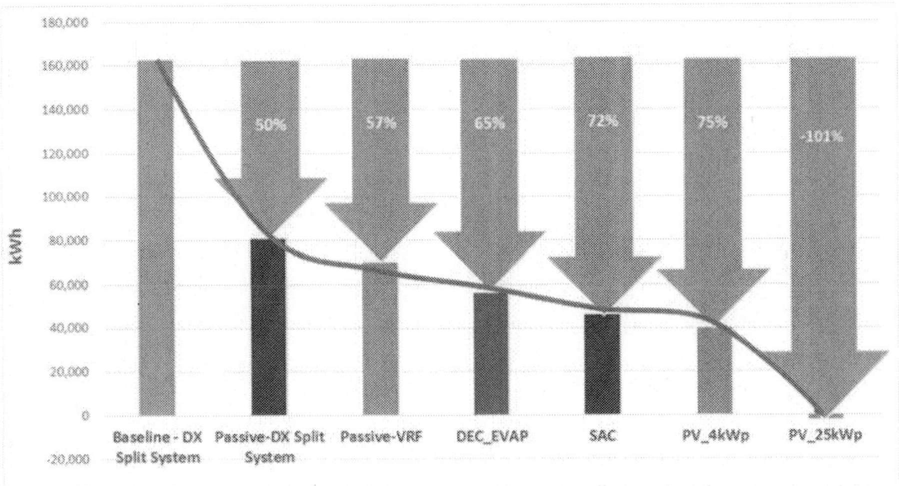

Figure 21. Total annual energy consumption with renewables (kWh).

5. Conclusion

This chapter presented the drivers, challenges and innovative technologies used to design the 'showcase' or the near zero energy home (nZEH) in Qatar (hot-humid climate). Several energy reduction strategies, using a defined energy hierarchy, are presented. The Be Lean or passive design measures are implemented comprehensively to reduce the needs for cooling. *This is* not limited to using high-performance multi-functional insulated façade, high performance glazing with *extensive* shading, the provision of day-lighting and energy efficient lighting without compromising the indoor environmental quality, *in order to* reduce the thermal loads. These measures contributed in reducing the cooling needs by 48%. The Be Clean strategies incorporated 'all-in-one' efficient cooling technologies that integrated desiccant cooling, integrated with indirect-direct evaporative cooling, absorption *cooling* and VRF system. The 5TR solar driven (50 m^2 solar collectors) absorption chiller is used to offset the cooling loads and the photovoltaic (4 kWp) is used to run auxiliaries and small power. The overall energy reduction is found to be 75%. In order to have a net zero energy building, it is necessary to upgrade the photovoltaic to 25 kWp, which means more than 200 m^2 of area is required which is not feasible. Although the nZEH is in the initial construction stage, it is anticipated that some minor changes into the design may occur. The monitoring results will be reported to the industry and community which will provide the pathway to the net zero energy building.

Author details

Esam Elsarrag* and Yousef Alhorr

*Address all correspondence to: e.elsarrag@gord.qa; elsarrag@hotmail.com

Gulf Organisation for Research and Development, Qatar Science and Technology Park, Doha, Qatar

References

[1] Alhorr Y, Eliskandarani E, Elsarrag E: Approaches to reducing carbon dioxide emissions in the built environment: Low carbon cities. International Journal of Sustainable Built Environment. 2014;3:167–178.

[2] Afshari A, Nikolopoulou C, Martin M: Life-cycle analysis of building retrofits at the urban scale — A case study in United Arab Emirates. Sustainability. 2014;6:453–473.

[3] IEA Energy Atlas, Qatar country profile. Available from: http://energyatlas.iea.org/?subject=1378539487 [Accessed: 2015-10-29]

[4] Elsarrag E, Alhorr Y: Modelling the thermal energy demand of a passive-house in the Gulf Region. International Journal of Sustainable Built Environment. 2012;1:1–16.

[5] IPCC AR4 WG3: Climate Change 2007: Mitigation of Climate Change. Contribution of Working Group III to the Fourth Assessment Report. Cambridge University Press, United Kingdom and New York, NY, USA; 2007. ISBN 978-0-521-88011-4 [Published: 978-0-521-70598-1]

[6] Alhorr Y, Elsarrag E: Climate change mitigation through energy benchmarking in the GCC green buildings codes. Buildings. 2015;5:700–714. DOI:10.3390/5020700. ISSN 2075-5309

[7] The UAE Solar Atlas [Internet]. 2013. http://www.mesia.com/wp-content/uploads/UAE-Solar-Resource-Atlas.pdf [Accessed: May 2016]

[8] Garde F, David M, Lenoir A, Ottenwelter E: Towards net zero energy in hot climate: Part 1, new tools and method. ASHRAE Transactions. 2011;117:1–8.

[9] Meijer F, Itard L, Sunikka-Blank M: Comparing European residential building stocks: performance renovation. Building Research & Information; 2009;37(5):533–551.

[10] US Department of Energy. A common definition for zero energy buildings. September 2015. Available from: http://energy.gov/sites/prod/files/2015/09/f26/A%20Common%20Definition%20for%20Zero%20Energy%20Buildings.pdf [Accessed: June 2016]

[11] Torcellini P, Pless S, Deru M: Zero energy buildings: A Critical Look at the Definition. National Renewable Energy Laboratory. In: Drury Crawley, U.S. Department of Energy. National Renewable Energy Laboratory report: NREL/CP-550-39833. 2006. Available from: http://www.nrel.gov/docs/fy06osti/39833.pdf

[12] Satori I, Napolitano A, Voss K: Net zero energy building. A consistent definition framework. Energy and Building. 2012;48:220–232

[13] Building Technologies Program. Planned Program Activities for 2008 – 2018. US Department of Energy. Energy Efficiency and Renewable Energy. 2008. Available from: http://apps1.eere.energy.gov/buildings/publications/pdfs/corporate/myp08complete.pdf [Accessed: 2016-10-04)

[14] EPBD recast, Directive 2010/31/EU of the European Parliament and of the Council of 19 May 2010 on the energy performance of buildings (recast). Official Journal of European Union [Internet]. 2010. Available from http://task40.iea-shc.org/objectivess [Accessed: 25-08-2015]

[15] Task 40. International Energy Agency- Solar Heating and Cooling. IEA-SHC., http://task40.iea-shc.org/objectives [Accessed: 2016-07-15]

[16] Marszal A, Heiselberg P: Zero energy building definition: A Literature Review. 2012. Available from: http://task40.iea-shc.org/data/sites/1/publications/T40A52-STA-Marzal-Report-2012-09.pdf2011 [Accessed: 2016-10-04]

[17] Torcellini P, Crawley D: Understanding zero-energy buildings. ASHRAE Journal. 2006;48:9:62–69

[18] Elsarrag E, Sahawneh F, Imbabi MS: The UAE Eco-Villa research and demonstration project. WRECXI: Abu Dhabi; 2010;25–30

[19] Elsarrag E, Alhorr Y: Using dynamic facade for indoor air quality, thermal comfort and energy efficient air conditioning. University of Khartoum Engineering Journal. 2011;1(2):63–66. ISSN 18586333

[20] Elsarrag E: An alternative system to supply treated fresh air for high rise buildings in humid climates and energy savings in UAE. Refrigeration, Heating and Air Conditioning. 2008;400:60–68. ISSN 02100665

[21] Utama A, Shabbir H: Indonesian residential high rise buildings. A life cycle energy assessment. Energy and Buildings. 2009;41:1263–1268. ISSN 0378-7788

[22] Ochoa C, Capeluto I: Strategic decision-making for intelligent buildings: Comparative impact of passive design strategies and active features in a hot climate. Building and Environment. 2008;43:1829–1839

[23] Cheung C: Energy-efficient envelope design for high-rise apartments. Energy and Buildings. 2005;37:37–48

[24] Niu J: Some significant environmental issues in high-rise residential building design in urban areas. Energy and Buildings. 2004;36:1259–1263.

[25] Priyadarsini R, Cheong KW, Wong NH: Enhancement of natural ventilation in high-rise residential buildings using stack system. Energy and Buildings. 2004;36:61–71. ISSN 0378-7788

[26] Lam JC, Yang L, Liu J: Development of passive design zones in China using bioclimatic approach. Energy Conversion and Management. 2006;47:746–762

[27] Gong X, Akashi Y, Sumiyoshi D: Optimization of passive design measures for residential buildings in different Chinese areas. Building and Environment. 2012;58:46–57

[28] Alawaji SH: Evaluation of solar energy research and its application in Saudi Arabia – 20 years of experience. Renewable and Sustainable Energy Reviews. 2001;5:59–77

[29] Hepbasli A, Alsuhaibani Z: A key review on present status and future directions of solar energy studies and application in Saudi Arabia. Renewable and Sustainable Energy Reviews. 2001;15:5021–5050

[30] Elsarrag E: Evaporation rate of a novel tilted solar liquid desiccant regeneration system. Solar Energy. 2008;82:663–668

[31] Elsarrag E, Magzoub E, Jain S: Design guidelines and performance study on a structured packed liquid desiccant air-conditioning system. ASHRAE HVAC&R Journal. DOI: 10.1080/10789669.2005.10391140

[32] Ssembatya M, Pokhrel MK, Reddy R: Simulation studies on performance of solar cooling system in UAE condition. Energy Procedia. 2014;48:1007–1016

[33] Deng S, Wang RZ, Dai YJ: How to evaluate performance of net zero energy building – A literature research. Energy. 2014;71:1–16.

[34] Still Cool. Construction week online. 2014. Available from: http://www.constructionweekonline.com/article-10057-still-cool/1/ [Accessed: 2015-10-30]

[35] GORD. http://www.gord.qa/ [Retrieved: 2014-12]

[36] Konstantinou T, Knaack U: An approach to integrate energy efficiency upgrade into refurbishment design process, applied in two case study buildings in Northern European Climate. Energy and Buildings. 2013;59:301–309.

[37] Elsarrag E, Aboulnaga M, Peacock A, Imbabi MS: Dynamic insulation for energy conservation and improved indoor air quality in hot humid climates. ASHRAE 5th Chapter Regional Conference (CRC); 01–03 November; Dubai, UAE (2006).

[38] Elsarrag E, Imbabi M: Evaluation of dynamic insulation for zone ventilation and air conditioning in the gulf region. ASHRAE Symposium on Sustainability and Green Buildings; 5 October, Kuwait (2009).

[39] Imbabi M: A passive–active dynamic insulation system for all climates. International Journal of Sustainable Built Environment. 2012;12;247–258.

[40] Elsarrag E, Alhorr Y, Imbabi M: Improving building fabric energy efficiency in hot-humid climates using dynamic insulation. Building Simulation. 2012;5:127–134. DOI 10.1007/s12273-012-0067-6.

[41] Siroky J, Oldewartel F, Cigler J, Privara S: Experimental analysis of model predictive control for an energy efficient building heating system. Applied Energy. 2011;88:3079–3087.